Topics in modern mathematics 2

Topics in modern mathematics 2

T. D. H. Baber

Ph.D., M.Sc., B.Sc., Dip. Ed.
Principal of Farnborough Technical College

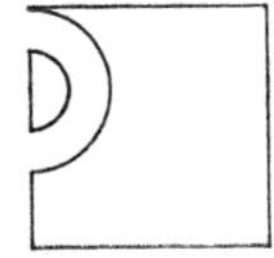

Pitman Publishing

First published 1973

SIR ISAAC PITMAN AND SONS LTD.
Pitman House, Parker Street, Kingsway, London, WC2B 5PB
P.O. Box 6038, Portal Street, Nairobi, Kenya

SIR ISAAC PITMAN (AUST.) PTY. LTD.
Pitman House, 158 Bouverie Street, Carlton, Victoria 3053, Australia

PITMAN PUBLISHING COMPANY S.A. LTD.
P.O. Box 11231, Johannesburg, S. Africa

PITMAN PUBLISHING CORPORATION
6 East 43rd Street, New York, N.Y. 10017, U.S.A.

SIR ISAAC PITMAN (CANADA) LTD.
495 Wellington St West, Toronto 135, Canada

THE COPP CLARK PUBLISHING COMPANY
517 Wellington St West, Toronto 135, Canada

Cased edition: ISBN 0 273 31680 X
Paperback edition: ISBN 0 273 31682 6

Made in Great Britain at the Pitman Press, Bath
G3-(T.374/1361: 75)

Preface

Over recent years, the approach to the teaching of mathematics has changed considerably in recognition of the need to provide a more effective programme of mathematical education in schools and colleges. There is general agreement that curricula and teaching methods require modernization. Reforms have been proposed by various organizations with the aims of promoting an early understanding of the basic structure of mathematics and of eliminating outmoded traditional material.

In various modern programmes, attention has been concentrated upon algebra, analysis and geometry treated in a more advanced and abstract way than formerly. The traditional approach, by which these subjects were compartmentalized, has been discarded and the present aim is to achieve a unified approach by which the relationships between these subjects are fully exploited and the underlying unity of mathematics is exhibited. It has been deemed desirable and practicable to introduce topics, formerly reserved for more advanced courses, at an earlier stage, e.g. the concepts and language of set theory are now widely recognized as providing an excellent medium for promoting the understanding and appreciation of a wide range of mathematical topics.

This volume presents a number of modern topics suitable not only for students who are proceeding to the study of mathematics, science or engineering but also as a programme of general education on the reasonable assumption that modern mathematics can make an important contribution in the field of liberal education. Whilst the presentation of these topics is non-traditional, students, whose earlier mathematical education has proceeded on traditional lines, should experience no handicap.

Volume I is based upon a series of lectures given by the author a few years ago to all undergraduates in the University of Malawi, irrespective of their specializations, to introduce them to modern mathematical thinking and teaching as a form of liberal education. The author makes no claim for comprehensiveness though the basic topics of mathematics and certain interesting and important applications of the subject have been included in this volume.

The two volumes, comprising this textbook, have not been written to cover any specific syllabuses. It is, however, claimed that they contain logically developed expositions of the more important topics of modern mathematics which have gained prominence in a wide range of examination syllabuses.

Volume 1 deals with the structure of the real number system and their representation on the number line. The study of inequalities, ordered pairs, relations and functions is developed in terms of set theory which is itself considered in an early chapter. A chapter on non-decimal arithmetic is followed by an account of digital computers and the elements of programming. This volume concludes with chapters on linear programming and an introduction to matrices and vectors. This volume is suitable for Vth and VIth formers in secondary schools, particularly for those who are studying modern mathematics syllabuses and also for students in Technical Colleges who are pursuing O.N.D./C. courses in Engineering and Science into which modern topics are being increasingly introduced.

Volume 2 contains subject matter of a more advanced standard including vector differentiation and integration, probability theory, Boolean algebras, and group theory. Since these topics are included in H.N.D./C. and degree courses in Mathematics, Science and Engineering, this volume is suitable for students pursuing such courses in Technical Colleges, Polytechnics and Universities.

In conclusion, I wish to thank Dr. Lee Peng Yee of the University of Malawi and Mr. R. W. Boxer for their helpful comments and suggestions.

T. D. H. Baber

Contents

6 *Boolean algebras*

7 *Residue classes*

8 *Groups*

1 Geometry of vectors

1.1 A Vector as an Equivalence Class

In Chapter 9 of volume 1, two- and three-dimensional vectors have been represented by (2×1) and (3×1) matrices respectively and various transformations in a plane have been studied in terms of such matrices. We now consider vectors from the geometrical point of view.

Consider the set of all straight lines of various directions and lengths, i.e. the set of directed line segments. The equivalence relation "has the same direction and length" partitions this set of line segments into equivalence classes, each class containing all those line segments which have the same length and direction.

Each equivalence class now defines a *vector*. If the directed line $\overline{\mathrm{AB}}$ is an element* of a particular equivalence class, this class may be said to represent the vector **AB**.

Two vectors **AB** and **CD** are equal, i.e. **AB** = **CD** if $\overline{\mathrm{AB}}$ and $\overline{\mathrm{CD}}$ have the same direction and equal lengths.

We have seen that two vectors may be combined or "added" to form a third vector by a parallelogram (or triangle) law. Let $\overline{\mathrm{AB}} = \overline{\mathrm{A'B'}}$, both directed lines being drawn from the vector class **AB**. Similarly let $\overline{\mathrm{BC}} = \overline{\mathrm{B'C'}}$; then we have†

$$\overline{\mathrm{AB}} + \overline{\mathrm{BC}} = \overline{\mathrm{AC}} \qquad \text{and} \qquad \overline{\mathrm{A'B'}} + \overline{\mathrm{B'C'}} = \overline{\mathrm{A'C'}}$$

* The bar indicates that $\overline{\mathrm{AB}}$ represents a displacement of magnitude AB in the direction from A towards B.

† The + sign here is used to denote *vector* addition.

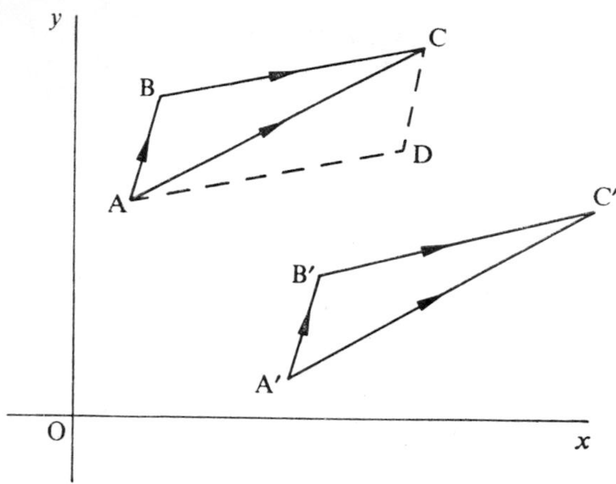

Figure 1.1

It is obvious that the triangles ABC, A′B′C′ are congruent and that corresponding sides are parallel. Therefore

$$\overline{\text{AC}} = \overline{\text{A}'\text{C}'}$$

so that both are drawn from the vector class **AC**.

It follows that if any pair of elements be drawn one from each of two vector classes, their sum will always be an element of a third unique vector class. The third vector is defined to be the *vector sum* of the first two vectors.

From the parallelogram in Fig. 1.1, we have

$$\overline{\text{AD}} + \overline{\text{DC}} = \overline{\text{AC}}$$

i.e.

$$\overline{\text{BC}} + \overline{\text{AB}} = \overline{\text{AC}} = \overline{\text{AB}} + \overline{\text{BC}}$$

so that vector addition is commutative.

If, in Fig. 1.1, $\overline{\text{AB}}$ is represented by $\begin{pmatrix} a \\ b \end{pmatrix}$ and $\overline{\text{BC}}$ by $\begin{pmatrix} c \\ d \end{pmatrix}$, then $\overline{\text{AC}}$ will be represented by $\begin{pmatrix} a+c \\ b+d \end{pmatrix}$.

Thus an isomorphism exists between (2×1) matrices under addition and vectors under a parallelogram law of addition.

1.2 Vectors and Scalars

A *vector* is a quantity which has both magnitude and direction, e.g. displacement, velocity, acceleration, force.

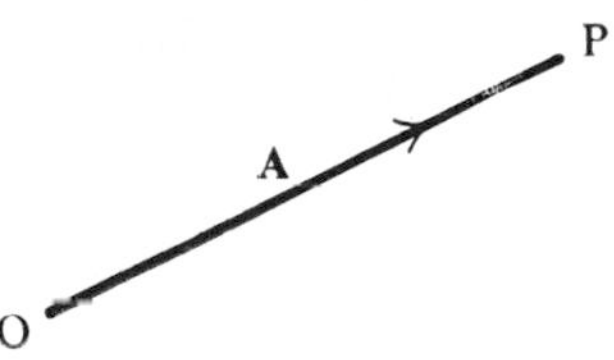

Figure 1.2

otation

he vector OP (Fig. 1.2) will be represented by **OP** and its magnitude by |**OP**|. It is often convenient, however, to represent the vector **OP** by **A** and s magnitude by |**A**| or simply A.

A *unit vector* is a vector of unit magnitude. Thus $\mathbf{A}/A$ is a unit vector aving the direction of **A** ($A \neq 0$).

A *scalar* is a quantity which has magnitude but no direction, e.g. any real umber, mass, length, time, temperature. Scalars will be denoted by ordinary tters like m, n.

.3 Vector Algebra

he familiar operations of addition, subtraction and multiplication in rdinary algebra may, by suitable definition, be extended to develop an gebra of vectors.

DEFINITION 1.1

wo vectors **A** and **B** are equal, i.e. $\mathbf{A} = \mathbf{B}$, if they have the same direction nd magnitude.

DEFINITION 1.2

he zero, null or neutral vector is a vector which is the neutral element for ector addition. It will clearly be a vector of zero magnitude and will be enoted by

$$\mathbf{0} \quad \text{and} \quad \mathbf{A} + \mathbf{0} = \mathbf{A}$$

DEFINITION 1.3

vector having a direction opposite to that of **A** but having the same magnide will be denoted by $-\mathbf{A}$.

ADDITION

Vectors are added by a parallelogram law and

$$\mathbf{A} + \mathbf{B} = \mathbf{B} + \mathbf{A} \qquad \text{and} \qquad \mathbf{A} + (-\mathbf{A}) = \mathbf{0}$$

The addition of vectors is associative, that is

$$\mathbf{A} + (\mathbf{B} + \mathbf{C}) = (\mathbf{A} + \mathbf{B}) + \mathbf{C}$$

In Fig. 1.3,

$$\mathbf{OP} + \mathbf{PQ} = \mathbf{OQ} = (\mathbf{A} + \mathbf{B}) \qquad \text{and} \qquad \mathbf{OQ} + \mathbf{QR} = \mathbf{OR} = \mathbf{D}$$

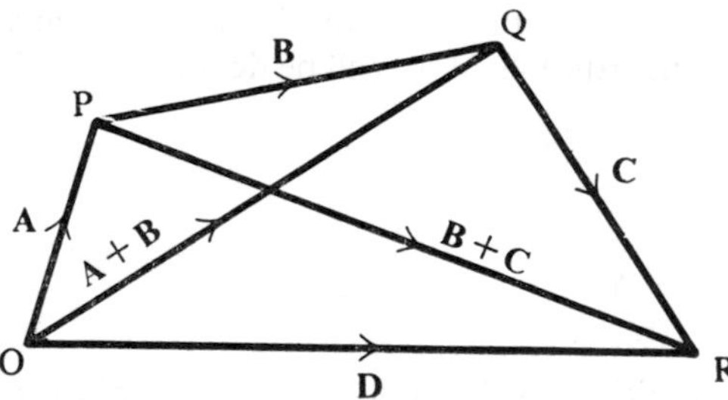

Figure 1.3

Therefore

$$(\mathbf{A} + \mathbf{B}) + \mathbf{C} = \mathbf{D}$$

Similarly,

$$\mathbf{PQ} + \mathbf{QR} = \mathbf{PR} = (\mathbf{B} + \mathbf{C}) \qquad \text{and} \qquad \mathbf{OP} + \mathbf{PR} = \mathbf{OR} = \mathbf{D}$$

Therefore

$$\mathbf{A} + (\mathbf{B} + \mathbf{C}) = \mathbf{D}$$

and the result follows.

SUBTRACTION

The difference of two vectors **A** and **B** is that vector which when added to **B** gives **A** (Fig. 1.4). This is equivalent to defining $\mathbf{A} - \mathbf{B}$ as the sum of **A** and $-\mathbf{B}$, i.e.

$$\mathbf{A} - \mathbf{B} = \mathbf{A} + (-\mathbf{B})$$

If $\mathbf{A} = \mathbf{B}$ then $\mathbf{A} - \mathbf{B} = 0$

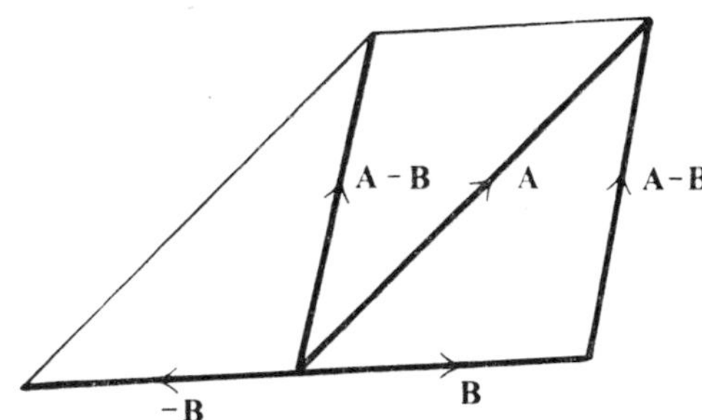

Figure 1.4

MULTIPLICATION BY A SCALAR

The product of a vector **A** by a scalar m is a vector $m\mathbf{A}$ whose magnitude is m times that of **A**. It has the same direction as **A** if $m > 0$ and has the opposite direction to **A** if $m < 0$. If $m = 0$, $m\mathbf{A}$ is a null vector.

LAWS OF VECTOR ALGEBRA

(1) $\mathbf{A} + \mathbf{B} = \mathbf{B} + \mathbf{A}$	Commutative Law of Addition
(2) $\mathbf{A} + (\mathbf{B} + \mathbf{C}) = (\mathbf{A} + \mathbf{B}) + \mathbf{C}$	Associative Law of Addition
(3) $m\mathbf{A} = \mathbf{A}m$	Commutative Law of Multiplication
(4) $m(n\mathbf{A}) = (mn)\mathbf{A}$	Associative Law of Multiplication
(5) $(m + n)\mathbf{A} = m\mathbf{A} + n\mathbf{A}$	Distributive Laws
(6) $m(\mathbf{A} + \mathbf{B}) = m\mathbf{A} + m\mathbf{B}$	

1.4 Position Vectors

For a given origin O, the position vectors of points A, B, C, . . . , with respect to O are **OA**, **OB**, **OC**, ... It will often be convenient to use the abbreviations **a**, **b**, **c**, . . . , respectively, for their position vectors.

1.5 The Ratio Theorem

We now find the vector equation of the straight line through two points A and B whose position vectors with respect to O are **a** and **b** respectively (Fig. 1.5).

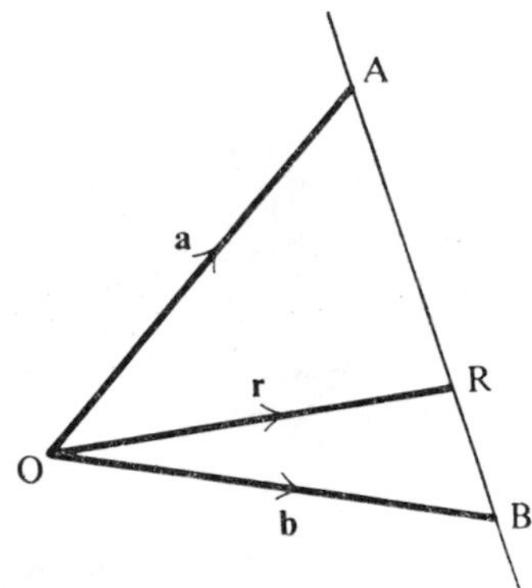

Figure 1.5

Let R be a point in AB such that $\mathbf{OR} = \mathbf{r}$.

$$\mathbf{AR} = \mathbf{r} - \mathbf{a} \qquad \text{and} \qquad \mathbf{AB} = \mathbf{b} - \mathbf{a}$$

Since **AR** and **AB** are collinear, $\mathbf{AR} = t\mathbf{AB}$, and

$$\mathbf{r} - \mathbf{a} = t(\mathbf{b} - \mathbf{a})$$

Thus the vector equation of AB with respect to O is

$$\mathbf{r} = \mathbf{a}(1 - t) + \mathbf{b}t$$

This equation may be written

$$\mathbf{r} = \lambda\mathbf{a} + \mu\mathbf{b} \tag{1.1}$$

where $\lambda + \mu = 1$.

$$\frac{\text{AR}}{\text{RB}} = \frac{\text{AR}}{\text{AB} - \text{AR}} = \frac{t\text{AB}}{(1 - t)\text{AB}} = \frac{t}{1 - t} = \frac{\mu}{\lambda}$$

Thus eqn (1.1) gives the position vector of a point R which divides AB in the ratio $\mu:\lambda$. (This ratio will be positive or negative according to whether R divides AB internally or externally.)

The symmetric form of this theorem may be obtained by noting that, since AR and RB are collinear, there are scalars m, n such that

$$m\mathbf{AR} = n\mathbf{RB}$$

i.e.

$$m(\mathbf{r} - \mathbf{a}) = n(\mathbf{b} - \mathbf{r})$$

$$\mathbf{r} = \frac{m\mathbf{a} + n\mathbf{b}}{m + n} \qquad (1.2)$$

This result may also be obtained by writing $\lambda = \dfrac{m}{m+n}$, $\mu = \dfrac{n}{m+n}$ in eqn (1.1).

Equation (1.2) may be written

$$(m + n)\mathbf{r} - m\mathbf{a} - n\mathbf{b} = 0$$

in which the sum of the coefficients of **r**, **a** and **b** is zero. If A, R, B are distinct collinear points, none of these coefficients is zero. Thus if A, R, B are distinct collinear points, numbers l, m, n exist, different from zero, such that

$$l\mathbf{r} + m\mathbf{a} + n\mathbf{b} = 0 \qquad l + m + n = 0 \qquad (1.3)$$

and conversely.

1.6 Centroids

The centroid of two particles of masses m_1, m_2 placed at A and B respectively is a point R on AB such that $AR/RB = m_2/m_1$.

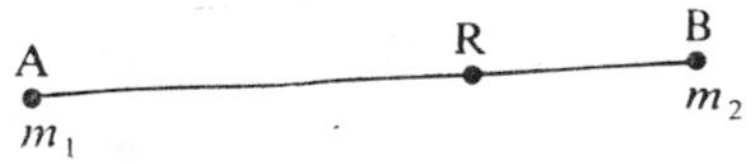

A R B
m_1 m_2

Figure 1.6

In applying the principles of mechanics, the two particles may usually be replaced by a single particle of mass $(m_1 + m_2)$ at R.

Using the symmetrical form of the ratio theorem, we have

$$(m_1 + m_2)\mathbf{r} = m_1\mathbf{a} + m_2\mathbf{b}$$

If there is a third particle of mass m_3 at C (Fig. 1.7), the centroid G of the three particles may be found as the centroid of mass $(m_1 + m_2)$ at R and

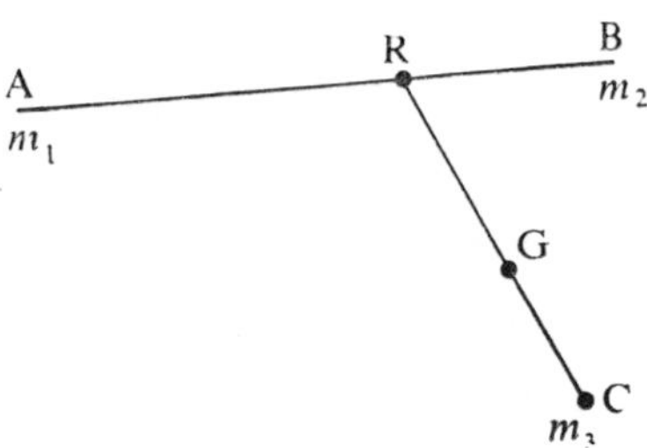

Figure 1.7

mass m_3 at C and will be given by

$$(m_1 + m_2 + m_3)\mathbf{g} = (m_1 + m_2)\mathbf{r} + m_3\mathbf{c} = m_1\mathbf{a} + m_2\mathbf{b} + m_3\mathbf{c}$$

G is clearly in the plane defined by the positions of the three particles but the origin O may be anywhere.

By continued application of the above procedure, it easily follows that the centroid G of n particles of masses $m_1, m_2, m_3, \ldots, m_n$ at points $P_1, P_2, P_3, \ldots, P_n$ respectively is given by

$$\left(\sum_{r=1}^{n} m_r\right)\mathbf{OG} = \sum_{r=1}^{n} m_r\mathbf{OP}_r \tag{1.4}$$

It is often convenient to refer to G as the centroid of points $P_1, P_2, \ldots, P_n$ with *associated numbers* $m_1, m_2, \ldots, m_n$ respectively. Negative masses or numbers cause no difficulty: for example, the centroid G of points A and B with associated numbers m_1 and $-m_2$ respectively $(m_1 > m_2 > 0)$ is given by

$$(m_1 - m_2)\mathbf{r} = m_1\mathbf{a} - m_2\mathbf{b}$$

and G divides AB externally such that $AG/GB = m_2/m_1$ [cf. eqn (1.2)].

The centroid G is determined uniquely by eqn (1.4) irrespective of the choice of O. If this were not the case, let G′ be the centroid as determined from another origin O′. Then

$$\left(\sum_{r=1}^{n} m_r\right)\mathbf{O'G'} = \sum_{r=1}^{n} m_r\mathbf{O'P}_r$$

Therefore

$$\left(\sum_{r=1}^{n} m_r\right)(\mathbf{OG} - \mathbf{O'G'}) = \sum_{r=1}^{n} m_r(\mathbf{OP}_r - \mathbf{O'P}_r) = \left(\sum_{r=1}^{n} m_r\right)\mathbf{OO'}$$

Since $\sum_{r=1}^{n} m_r \neq 0$, $\mathbf{OG} - \mathbf{O'G'} = \mathbf{OO'}$, then

$$\mathbf{OG} = \mathbf{O'G'} + \mathbf{OO'} = \mathbf{OG'}$$

The ratio theorem, eqn (1.2), and its extension to centroids, eqn (1.4), are most useful in proving many theorems in geometry.

EXAMPLE 1.1
Show that the medians of a triangle are concurrent.

Solution Let G be the centroid of equal masses placed at the vertices A, B,

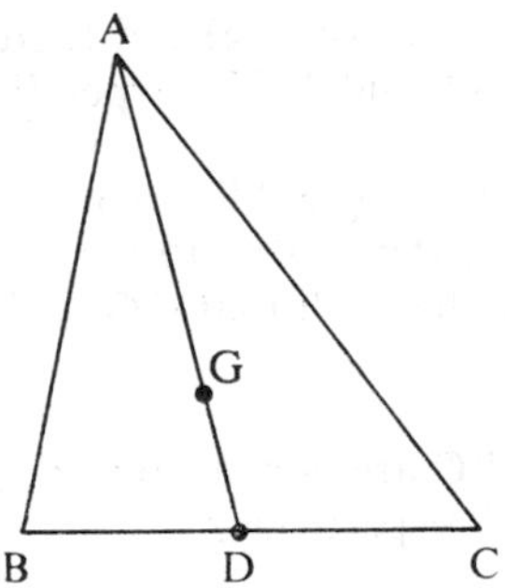

Figure 1.8

C of the triangle and let D be the mid-point of BC (Fig. 1.8).

$$\mathbf{g} = \tfrac{1}{3}(\mathbf{a} + \mathbf{b} + \mathbf{c}) = \tfrac{1}{3}\mathbf{a} + \tfrac{2}{3}.\tfrac{1}{2}(\mathbf{b} + \mathbf{c}) = \tfrac{1}{3}\mathbf{a} + \tfrac{2}{3}\mathbf{d}$$

By the ratio theorem, G lies on the median AD such that AG/GD = 2/1.

Similarly the centroid G lies at the points of trisection of the other two medians. Hence the medians are concurrent at G which is called the *centroid* of the triangle.

EXAMPLE 1.2

Show that the altitudes of a triangle are concurrent.

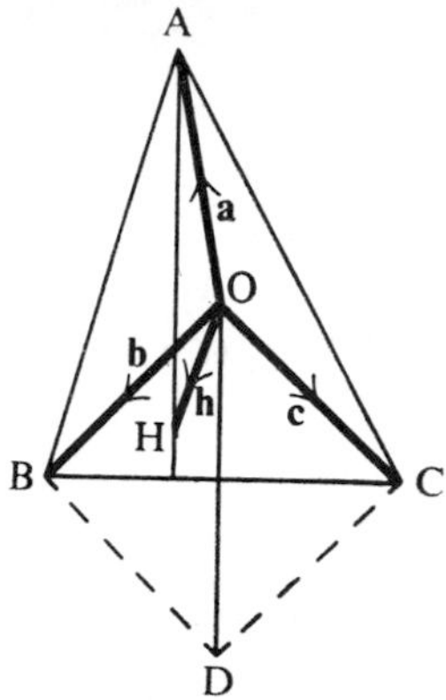

Figure 1.9

Solution Let the circumcentre of the triangle ABC in Fig. 1.9 be chosen as origin O and let $\mathbf{h} = \mathbf{a} + \mathbf{b} + \mathbf{c}$. Since $|\mathbf{a}| = |\mathbf{b}| = |\mathbf{c}|$, the parallelogram OCDB, having the vectors **b** and **c** as a pair of sides, is a rhombus.

Therefore, OD is perpendicular to BC.

Therefore, vector $(\mathbf{b} + \mathbf{c})$ is perpendicular to vector $(\mathbf{b} - \mathbf{c})$ and therefore Vector $(\mathbf{h} - \mathbf{a})$ is perpendicular to vector $(\mathbf{b} - \mathbf{c})$, and AH is perpendicular to BC.

Similarly, H is on the other two altitudes of the triangle ABC, so that the three altitudes are concurrent at H which is called the *orthocentre*.

Also since $\mathbf{g} = \frac{1}{3}(\mathbf{a} + \mathbf{b} + \mathbf{c}) = \frac{1}{3}\mathbf{h}$, G lies on OH such that $OG = \frac{1}{3}OH$.

EXAMPLE 1.3

Show that the internal bisectors of a triangle ABC are concurrent at I, the centroid of masses sin A, sin B, sin C at A, B, C respectively.

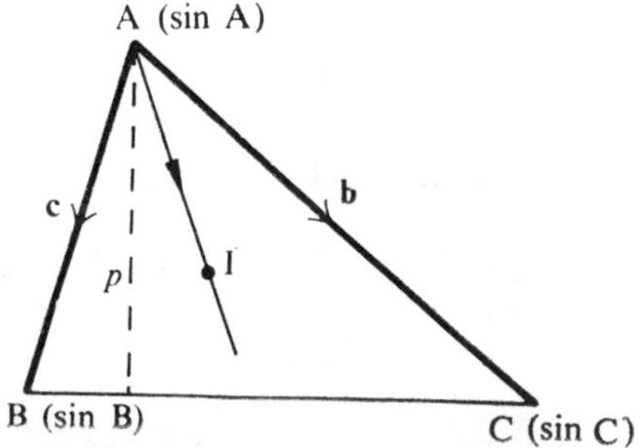

Figure 1.10

Solution Take A as origin. Then, if $\mathbf{AC} = \mathbf{b}$ and $\mathbf{AB} = \mathbf{c}$, from Fig. 1.10,

$$(\sin A + \sin B + \sin C)\mathbf{AI} = \mathbf{c} \sin B + \mathbf{b} \sin C$$

$$= \frac{\mathbf{c}}{c} c \sin B + \frac{\mathbf{b}}{b} b \sin C$$

$$= p\left(\frac{\mathbf{c}}{c} + \frac{\mathbf{b}}{b}\right)$$

where p is the altitude of the triangle through A.

Since $\mathbf{c}/c$ and $\mathbf{b}/b$ are unit vectors in the directions AB and AC respectively, it follows that $(\mathbf{c}/c) + (\mathbf{b}/b)$ is a vector in the direction of the internal bisector of the angle A. Thus **AI** has the direction of this bisector, so that the centroid I lies on this bisector, and similarly lies on the other two internal bisectors.

Therefore, the internal bisectors of a triangle are concurrent at I, the *incentre*.

EXAMPLE 1.4

Prove that the straight lines joining the vertices of a tetrahedron to the centroids of the opposite faces (the medians of the tetrahedron) are concurrent.

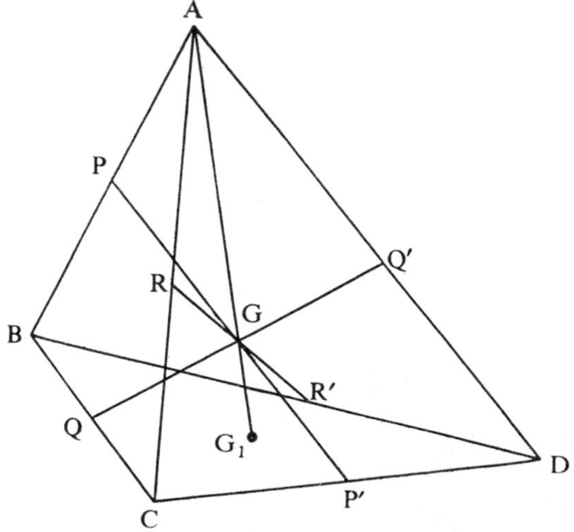

Figure 1.11

Solution Let G be the centroid of unit masses placed at the vertices A, B, C, D of a tetrahedron (Fig. 1.11). Let P, P′ be the mid-points of AB, CD respectively. Also let Q, Q′ be those of BC, AD, and R, R′ those of AC, BD respectively. Let G_1 be the centroid of triangle BCD. Then

$$\mathbf{g} = \tfrac{1}{4}(\mathbf{a} + \mathbf{b} + \mathbf{c} + \mathbf{d}) = \tfrac{1}{2}[\tfrac{1}{2}(\mathbf{a} + \mathbf{b}) + \tfrac{1}{2}(\mathbf{c} + \mathbf{d})] = \tfrac{1}{2}(\mathbf{p} + \mathbf{p}')$$

Similarly $\mathbf{g} = \tfrac{1}{2}(\mathbf{q} + \mathbf{q}') = \tfrac{1}{2}(\mathbf{r} + \mathbf{r}')$. It follows that G is the point of concurrence of the lines PP′, QQ′, RR′ which join the mid-points of opposite edges of the tetrahedron. Also

$$\mathbf{g} = \tfrac{1}{4}\mathbf{a} + \tfrac{3}{4}.\tfrac{1}{3}(\mathbf{b} + \mathbf{c} + \mathbf{d}) = \tfrac{1}{4}\mathbf{a} + \tfrac{3}{4}\mathbf{g}_1$$

Therefore G lies on AG_1 and divides AG_1 so that $AG:GG_1 = 3:1$. Thus G is also the point of concurrence of the four medians of the tetrahedron so that seven lines meet at G.

EXAMPLE 1.5

A transversal cuts the sides AB, BC, CA of the triangle ABC in the points D, E, F respectively. Show that the product of the ratios in which D, E, F divide these sides is -1 (Menelaus' Theorem).

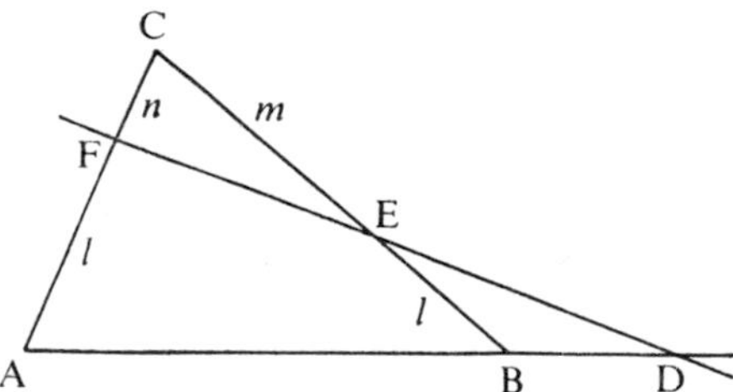

Figure 1.12

Solution Let E divide BC in the ratio $l:m$ (Fig. 1.12) and let F divide CA in the ratio $n:l$ so that

$$(l + m)\mathbf{e} = l\mathbf{c} + m\mathbf{b} \qquad \text{and} \qquad (n + l)\mathbf{f} = n\mathbf{a} + l\mathbf{c}$$

By subtracting, eliminate **c** so that

$$\frac{(l + m)\mathbf{e} - (n + l)\mathbf{f}}{m - n} = \frac{m\mathbf{b} - n\mathbf{a}}{m - n}$$

By the ratio theorem, eqn. (1.1), each of these fractions is equal to the position vector **d** since D lies in both FE and AB.

But clearly, from $(m\mathbf{b} - n\mathbf{a})/(m - n) = \mathbf{d}$, D divides AB in the ratio $m{:}-n$. Therefore

Product of the three ratios is $\dfrac{l}{m} \times \dfrac{n}{l} \times \dfrac{-m}{n} = -1.$

EXERCISES 1.1

1. Prove that the line joining the mid-points of two sides of a triangle is parallel to the third and has one half of its length.
2. If O is any point inside triangle ABC and P, Q, R are the mid-points of its sides, show that

   $$\mathbf{OA} + \mathbf{OB} + \mathbf{OC} = \mathbf{OP} + \mathbf{OQ} + \mathbf{OR}$$

   Show that the result also holds if O lies outside triangle ABC.
3. The mid-points of the consecutive sides of any quadrilateral (skew or otherwise) are joined. Show that the resulting quadrilateral is a parallelogram.
4. Show that the orthocentre of a triangle ABC coincides with the centroid of masses tan A, tan B, tan C at A, B, C respectively.
5. Show that the circumcentre of triangle ABC coincides with the centroid of masses sin 2A, sin 2B, sin 2C at A, B, C respectively.
6. If G, G′ are respectively the centroids of triangles ABC, A′B′C′ show that

   $$3\mathbf{GG'} = \mathbf{AA'} + \mathbf{BB'} + \mathbf{CC'}$$

7. P and Q are the mid-points of the sides AB, BC respectively of the parallelogram ABCD. Show that DP and DQ trisect the line AC. Prove also that AC passes through a point of trisection of DP and DQ.
8. Let D be the point on the side BC of triangle ABC such that BD/DC $= n/m$ and let R divide AD such that AR/RD $= (m + n)/l$. Show that

   $$\mathbf{r} = (l\mathbf{a} + m\mathbf{b} + n\mathbf{c})/(l + m + n)$$

   If D, E, F are points on BC, CA, AB respectively such that AD, BE, CF are concurrent at R, deduce that

   $$\frac{BD}{DC}\cdot\frac{CE}{EA}\cdot\frac{AF}{FB} = 1 \qquad \text{(Ceva's theorem)}$$

1.7 Rectangular Unit Vectors

Using a right-handed system of axes (Fig. 1.13), unit vectors along Ox, Oy, Oz will be denoted by **i**, **j**, **k** respectively.

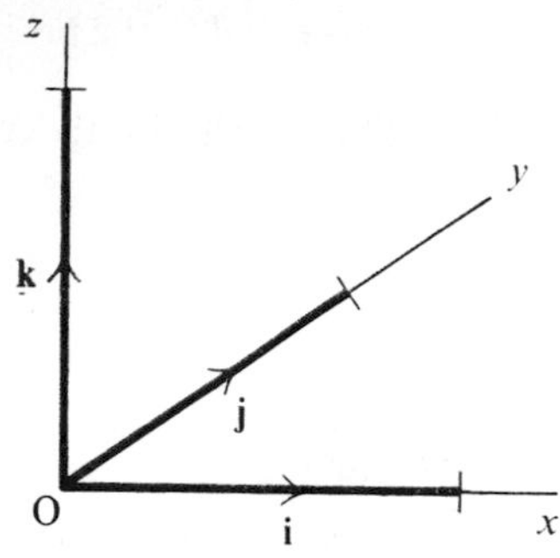

Figure 1.13

1.8 Components of a Vector

Let a vector **A** be localized at O (Fig. 1.14) and let the coordinates of the terminal point of **A** with respect to rectangular axes through O be (a_1, a_2, a_3).

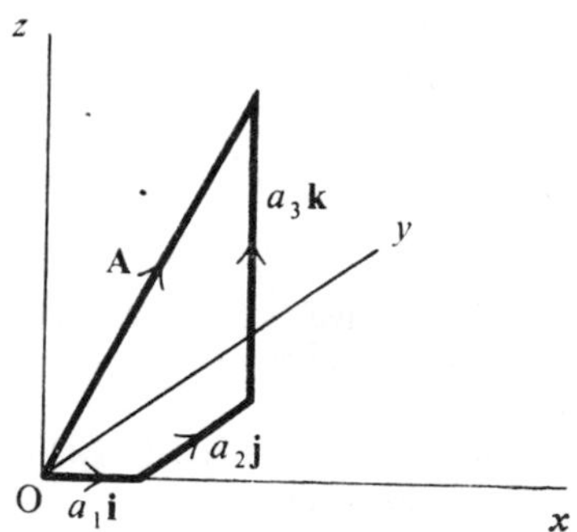

Figure 1.14

The vectors $a_1\mathbf{i}$, $a_2\mathbf{j}$, $a_3\mathbf{k}$ are called the *rectangular component vectors* of **A** with respect to the axes and a_1, a_2, a_3 are called its *rectangular components* with respect to the axes.

$$\mathbf{A} = a_1\mathbf{i} + a_2\mathbf{j} + a_3\mathbf{k}$$

and the magnitude of **A** is $A = \sqrt{(a_1^2 + a_2^2 + a_3^2)}$.

If **r** is the position vector of the point (x, y, z) with respect to O,

$$\mathbf{r} = x\mathbf{i} + y\mathbf{j} + z\mathbf{k} \qquad \text{and} \qquad r = \sqrt{(x^2 + y^2 + z^2)}$$

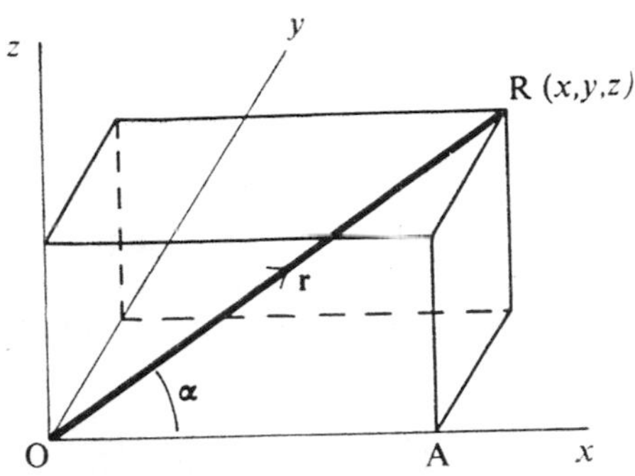

Figure 1.15

Let the vector **r** make angles α, β, γ with the positive directions of the coordinate axes Ox, Oy, Oz respectively (Fig. 1.15).

From the right-angled triangle OAR,

$$\cos\alpha = \mathrm{OA}/\mathrm{OR} = x/r$$

Similarly $\cos\beta = y/r$ and $\cos\gamma = z/r$. Therefore

$$\cos^2\alpha + \cos^2\beta + \cos^2\gamma = (x^2 + y^2 + z^2)/r^2 = 1$$

$\cos\alpha$, $\cos\beta$, $\cos\gamma$ are called the *direction cosines* of the vector **OR**.

Let **a** and **b** be two non-collinear vectors localized at O (Fig. 1.16). Let **r** be the vector **OR** in the plane determined by **a** and **b**. Through R draw BR, AR

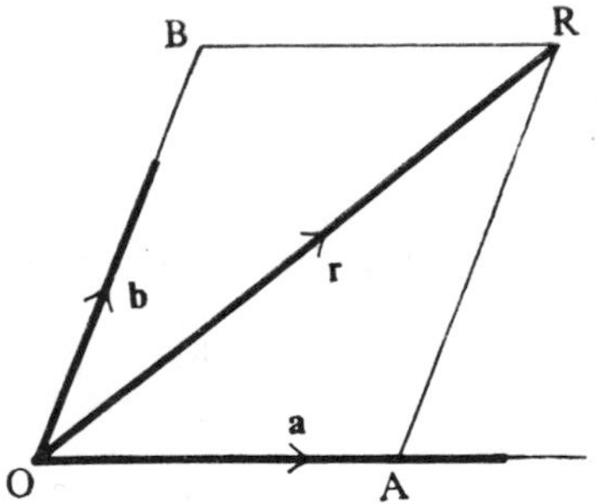

Figure 1.16

parallel to **a**, **b** respectively then

$$\mathbf{r} = \mathbf{OA} + \mathbf{OB} = x\mathbf{a} + y\mathbf{b}$$

where x, y are suitable scalars.

$x\mathbf{a}$ and $y\mathbf{b}$ are said to the *components* of **r** in the directions of **a** and **b** respectively and **a** and **b** are called *base vectors* in the plane.

Similarly a vector **r** in three-dimensional space may be expressed in terms of three non-coplanar base vectors **a**, **b**, **c** (Fig. 1.17).

$$\mathbf{r} = \mathbf{OA} + \mathbf{OB} + \mathbf{OC} = x\mathbf{a} + y\mathbf{b} + z\mathbf{c}$$

where x, y, z are suitable scalars.

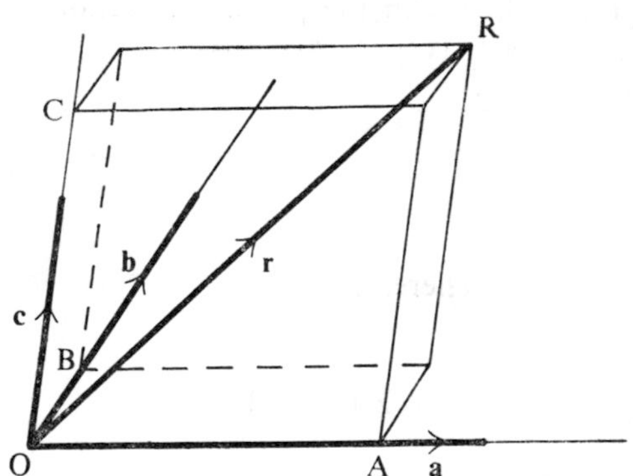

Figure 1.17

If $x\mathbf{a} + y\mathbf{b} + z\mathbf{c} = 0$, then $x = y = z = 0$, for suppose $x \neq 0$, then

$$\mathbf{a} = -\frac{y}{x}\mathbf{b} - \frac{z}{x}\mathbf{c}$$

so that **a** is a vector in the plane of **b** and **c**, contrary to hypothesis. Therefore $x = 0$ and similarly $y = z = 0$.

It follows that if $x_1\mathbf{a} + y_1\mathbf{b} + z_1\mathbf{c} = x_2\mathbf{a} + y_2\mathbf{b} + z_2\mathbf{c}$, so that

$$(x_1 - x_2)\mathbf{a} + (y_1 - y_2)\mathbf{b} + (z_1 - z_2)\mathbf{c} = 0$$

then $x_1 - x_2 = 0$ and $x_1 = x_2$. Similarly $y_1 = y_2$, $z_1 = z_2$.

1.9 Equation of the Straight Line through a Given Point, Parallel to a Given Vector

Let the required line pass through point A to be parallel to vector **b**. Let R be a point on the line (Fig. 1.18). Then

$$\mathbf{AR} = t\mathbf{b}$$
$$\mathbf{r} = \mathbf{OA} + \mathbf{AR} = \mathbf{a} + t\mathbf{b}$$

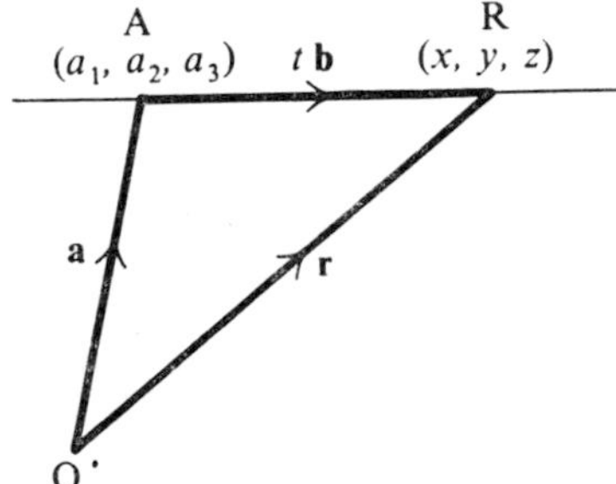

Figure 1.18

If R and A are respectively the points (x, y, z) and (a_1, a_2, a_3) and **b** has the components (b_1, b_2, b_3) then

$$x\mathbf{i} + y\mathbf{j} + z\mathbf{k} = (a_1\mathbf{i} + a_2\mathbf{j} + a_3\mathbf{k}) + t(b_1\mathbf{i} + b_2\mathbf{j} + b_3\mathbf{k})$$

Therefore

$$x = a_1 + tb_1 \qquad \frac{x - a_1}{b_1} = t$$

Thus the Cartesian equation of the line is

$$\frac{x - a_1}{b_1} = \frac{y - a_2}{b_2} = \frac{z - a_3}{b_3} = t$$

1.10 Equation of the Straight Line through Two Given Points

A and B are the two points (Fig. 1.19).

$$\mathbf{AB} = \mathbf{b} - \mathbf{a}$$

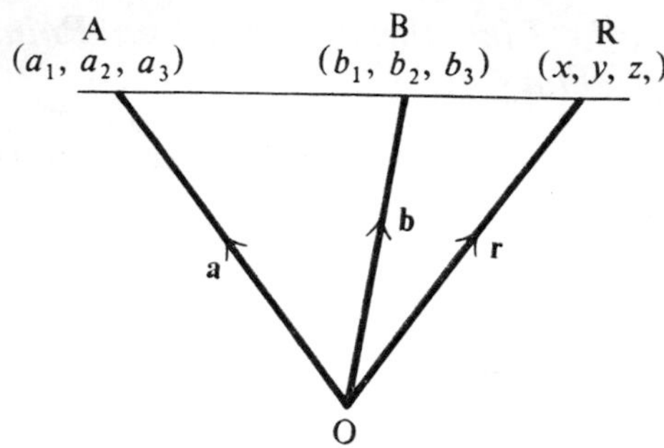

Figure 1.19

Therefore

$$\mathbf{AR} = t(\mathbf{b} - \mathbf{a})$$

$$\mathbf{r} = \mathbf{OA} + \mathbf{AR} = \mathbf{a} + t(\mathbf{b} - \mathbf{a})$$

i.e.

$$\mathbf{r} = (1 - t)\mathbf{a} + t\mathbf{b}$$

$$(x\mathbf{i} + y\mathbf{j} + z\mathbf{k}) = (a_1\mathbf{i} + a_2\mathbf{j} + a_3\mathbf{k}) + t[(b_1 - a_1)\mathbf{i} + (b_2 - a_2)\mathbf{j} + (b_3 - a_3)\mathbf{k}]$$

so that the Cartesian equation of the line is

$$\frac{x - a_1}{b_1 - a_1} = \frac{y - a_2}{b_2 - a_2} = \frac{z - a_3}{b_3 - a_3} = t$$

1.11 Equations of the Bisectors of the Angles between Two Unit Vectors Localized at a Given Point

Let R be a point on the internal bisector between the unit vectors **a** and **b** (Fig. 1.20). Let B′R and A′R be drawn parallel to **a** and **b** respectively.

$$|\mathbf{OA'}| = |\mathbf{A'R}| = |\mathbf{OB'}| \quad \text{so that} \quad \mathbf{OA'} = t\mathbf{a} \quad \text{and} \quad \mathbf{A'R} = t\mathbf{b}$$

But

$$\mathbf{r} = \mathbf{OA'} + \mathbf{A'R}$$

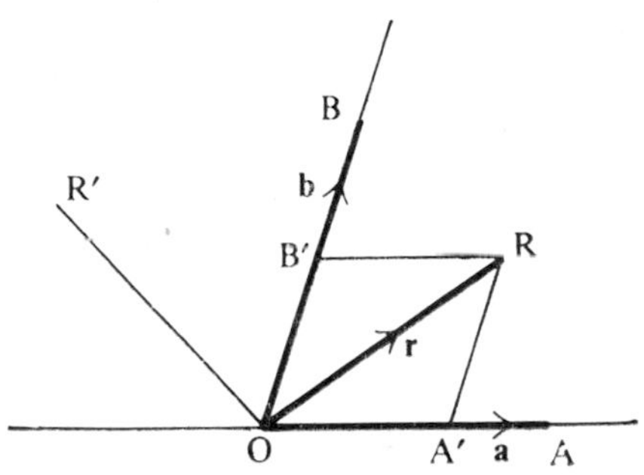

Figure 1.20

nd therefore

$$\mathbf{r} = t(\mathbf{a} + \mathbf{b})$$

imilarly the equation of the external bisector CR′ is $\mathbf{r} = t(\mathbf{a} - \mathbf{b})$. If **a** and **b** re not unit vectors, the equations of the two bisectors are

$$\mathbf{r} = t\left(\frac{\mathbf{a}}{a} \pm \frac{\mathbf{b}}{b}\right)$$

XAMPLE 1.6

R, the internal bisector of the angle BAC of triangle ABC, meets BC in R. how that BR/RC = c/b.

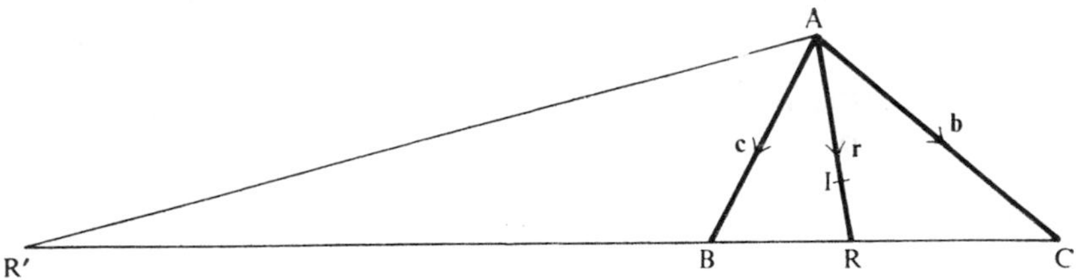

Figure 1.21

olution Taking A as origin, the equation of the internal bisector of angle AC is

$$\mathbf{r} = t\left(\frac{\mathbf{c}}{c} + \frac{\mathbf{b}}{b}\right)$$

If the scalar t is given the particular value $bc/(b + c)$, then

$$\mathbf{r} = \frac{b\mathbf{c} + c\mathbf{b}}{b + c}$$

is a point on AR. By the ratio theorem, it must be a point on BC and i therefore the point R. Thus R is the centroid of B and C with associate numbers b and c respectively. Thus BR/RC $= c/b$.

Similarly, since the *external* bisector of angle BAC is

$$\mathbf{r} = t\left(\frac{\mathbf{c}}{c} - \frac{\mathbf{b}}{b}\right)$$

$(b\mathbf{c} - c\mathbf{b})/(b - c)$ is a point R′ on it which divides CB externally such tha BR′/R′C $= c/b$.

EXAMPLE 1.7

Show that the internal bisectors of the angles of a triangle are concurrent

Solution The centroid of the points A, B, C with associated numbers a, b, respectively is a point I on AR (Fig. 1.21) such that AI/IR $= (b + c)/a$.

Similarly the centroid I must be on the internal bisectors of the angle ABC, ACB.

Therefore the three internal bisectors are concurrent at I.

Similarly, it may be shown that the internal bisector of the angle BAC an the external bisectors of the other two angles are concurrent at the centroi of the points A, B, C with associated numbers a, $-b$, $-c$ respectively.

1.12 Equation of the Plane through a Given Point, Paralle to Two Given Vectors

Let A be the given point. Let R be a point in the plane (Fig. 1.22), and **b** anc **c** the vectors. Since **AR** is parallel to **b** and **c**,

$$\mathbf{AR} = s\mathbf{b} + t\mathbf{c} \qquad \text{and} \qquad \mathbf{OR} = \mathbf{OA} + \mathbf{AR}$$

Therefore

$$\mathbf{r} = \mathbf{a} + s\mathbf{b} + t\mathbf{c}$$

As R moves in the plane, the scalars s and t take various values.

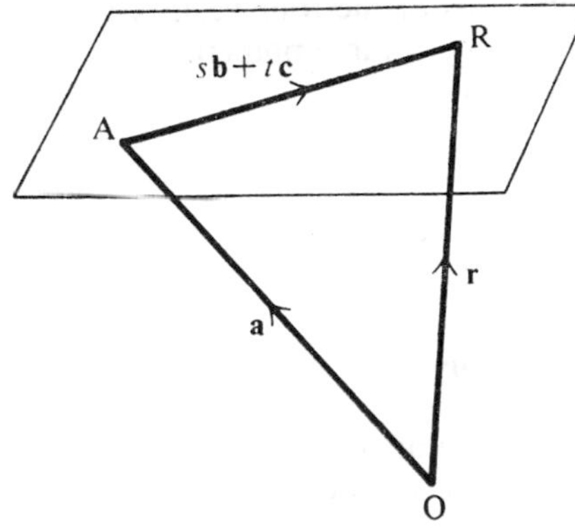

Figure 1.22

1.13 Equation of the Plane through Three Given Points

The required plane passes through A, B, C (Fig. 1.23) and is parallel to the vectors $\mathbf{AB} = \mathbf{b} - \mathbf{a}$ and $\mathbf{AC} = \mathbf{c} - \mathbf{a}$ so that the required equation is

$$\mathbf{r} = \mathbf{a} + s(\mathbf{b} - \mathbf{a}) + t(\mathbf{c} - \mathbf{a}) = (1 - s - t)\mathbf{a} + s\mathbf{b} + t\mathbf{c}$$

(It is assumed that the three vectors **a, b, c** are not coplanar.)
Thus four points R, A, B, C are coplanar if

$$(1 - s - t)\mathbf{a} + s\mathbf{b} + t\mathbf{c} - \mathbf{r} = 0$$

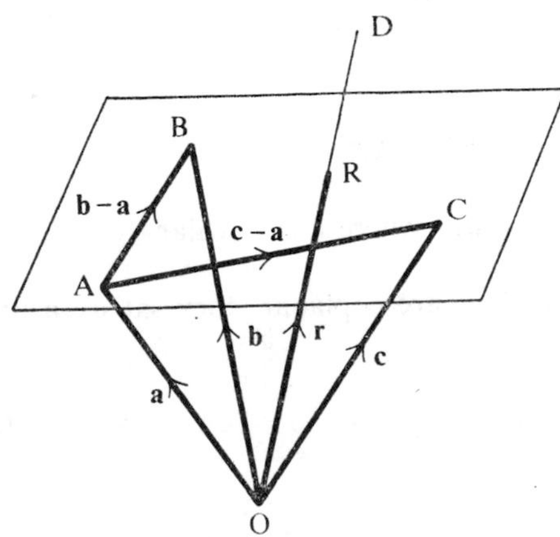

Figure 1.23

In this equation, the sum of the coefficients of the four vectors is zero (cf. the condition, eqn (1.3), for the collinearity of three points).

This result may be written in the equivalent form

$$l\mathbf{a} + m\mathbf{b} + n\mathbf{c} + p\mathbf{r} = 0 \tag{1.5}$$

where $l + m + n + p = 0$, and l, m, n, p are not all zero.

1.14 Linear Dependence of Vectors

(i) Let **a**, **b**, **c** be three distinct* coplanar vectors (Fig. 1.24). Let OC cut AB in R. Then by eqn (1.3), scalars l, m, n exist, each different from zero, such that

$$l\mathbf{a} + m\mathbf{b} + n\mathbf{r} = 0 \qquad l + m + n = 0$$

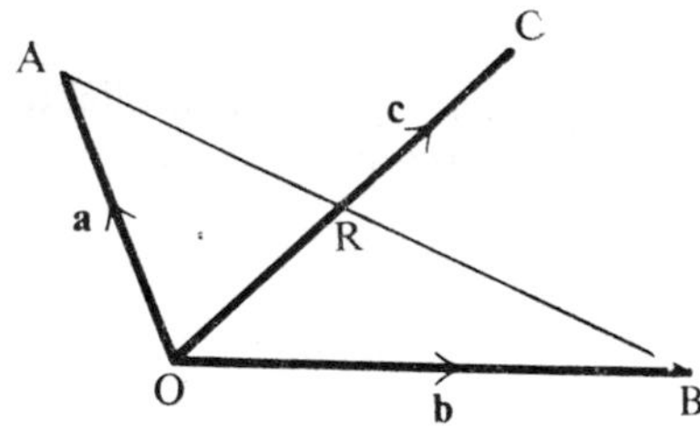

Figure 1.24

But $\mathbf{c} = t\mathbf{r}$ where t is a scalar and therefore

$$\alpha\mathbf{a} + \beta\mathbf{b} + \gamma\mathbf{c} = 0 \qquad \alpha, \beta, \gamma \neq 0$$

In the special case when **OC** is parallel to **AB**,

$$\mathbf{OC} = \lambda\mathbf{AB} \quad \text{so that} \quad \mathbf{c} = \lambda(\mathbf{b} - \mathbf{a}) \quad \text{where } \lambda \text{ is a scalar}$$

Thus when three distinct vectors **a**, **b**, **c** are coplanar, they satisfy a linear relationship of the form

$$\alpha\mathbf{a} + \beta\mathbf{b} + \gamma\mathbf{c} = 0 \tag{1.6}$$

and are said to be *linearly dependent*.

* The word "distinct" implies that no two vectors have the same direction.

(ii) Let **a**, **b**, **c**, **d** be four distinct vectors in space, of which no three lie in the same plane.

Let OD meet the plane containing A, B, C in R (Fig. 1.23). Then by eqn (1.5),

$$l\mathbf{a} + m\mathbf{b} + n\mathbf{c} + p\mathbf{r} = 0 \qquad l + m + n + p = 0$$

But $\mathbf{d} = t\mathbf{r}$ and therefore

$$\alpha\mathbf{a} + \beta\mathbf{b} + \gamma\mathbf{c} + \delta\mathbf{d} = 0 \tag{1.7}$$

α, β, γ, δ are not all zero but some may be zero.

In the particular case, when **OD** is parallel to the plane ABC,

$$\mathbf{d} = \lambda(\mathbf{b} - \mathbf{a}) + \mu(\mathbf{c} - \mathbf{a})$$

which is of the same form as eqn (1.7).

When four distinct vectors satisfy an equation of this form, they are said to be *linearly dependent*. Whilst three vectors in three dimensional space may be linearly independent, four vectors must be linearly dependent.

EXERCISES 1.2

1. **A** and **B** are given vectors. Show that

   (i) $|\mathbf{A} + \mathbf{B}| \leqslant |\mathbf{A}| + |\mathbf{B}|$ (ii) $|\mathbf{A} - \mathbf{B}| \geqslant |\mathbf{A}| - |\mathbf{B}|$

2. If P and Q are the points (x_1, y_1, z_1) and (x_2, y_2, z_2) respectively, find the magnitude of the vector **PQ**. If P is the point (2, 3, −1) and Q the point (4, −3, 2) show that **PQ** is the vector $2\mathbf{i} - 6\mathbf{j} + 3\mathbf{k}$ having a magnitude 7.
3. Show that equations of the straight line through the points (0, −2, 3) and (1, −2, 1) are $2x + z = 3$; $y = z$. Show that this straight line meets the plane determined by the origin and the points (2, 4, 1) and (4, 0, 2) in the point (6/5, 2, 3/5).
4. Show that the centroid of A (−2, 2, −1), B (2, −1, 3), C (−2, 4, 1), D (1, 2, 3) with associated numbers 1, 2, 3 and 4 respectively is (0, 2, 2).
5. Show that the vectors $\mathbf{A} = -\mathbf{i} + 3\mathbf{j} + 4\mathbf{k}$, $\mathbf{B} = 3\mathbf{i} + \mathbf{j} - 2\mathbf{k}$, $\mathbf{C} = 4\mathbf{i} - 2\mathbf{j} - 6\mathbf{k}$ are coplanar and show that the lengths of the medians of the triangle ABC are $\frac{1}{2}\sqrt{386}$, $\frac{1}{2}\sqrt{14}$, $\sqrt{74}$.
6. With distances measured in nautical miles and speeds in knots, three ships are observed from a coastguard station at half-hour intervals. They have the following distance (**s**) and velocity (**v**) vectors:

   $\mathbf{s}_1 = 2\mathbf{i} + 6\mathbf{j}$ and $\mathbf{v}_1 = 5\mathbf{i} + 4\mathbf{j}$ at 12 00
   $\mathbf{s}_2 = 6\mathbf{i} + 9\mathbf{j}$ and $\mathbf{v}_2 = 4\mathbf{i} + 3\mathbf{j}$ at 12 30
   $\mathbf{s}_3 = 11\mathbf{i} + 6\mathbf{j}$ and $\mathbf{v}_3 = 2\mathbf{i} + 7\mathbf{j}$ at 13 00

Prove that if the ships continue with the same velocities, two of them will collide, and find the time of collision. If at that instant the third ship changes course and then proceeds directly to the scene of collision at its original speed, find at what time it will arrive (*U.L.A-level*).
[Time of collision 14 20; $2\sqrt{29}$ minutes after the collision]

7. If the position vectors of points $P_1, P_2, \ldots, P_n$ with respect to an origin O are $\mathbf{r}_1, \mathbf{r}_2, \ldots, \mathbf{r}_n$ respectively and scalars $k_1, k_2, \ldots, k_n$ exist such that

$$k_1\mathbf{r}_1 + k_2\mathbf{r}_2 + \cdots + k_n\mathbf{r}_n = 0$$

then show that this result will be independent of the origin if, and only if, $k_1 + k_2 + \cdots + k_n = 0$.

8. The position vectors of three points A, B, C are respectively

$$\mathbf{a} = 2\mathbf{i} - 4\mathbf{j} - 3\mathbf{k} \qquad \mathbf{b} = -\mathbf{i} + 2\mathbf{j} + 2\mathbf{k} \qquad \mathbf{c} = 3\mathbf{i} - 2\mathbf{j} + \mathbf{k}$$

Express the vector $\mathbf{d} = -5\mathbf{i} + 2\mathbf{j} - \mathbf{k}$ as a linear function of **a**, **b**, **c**.
What point on **d** is in the plane ABC?
[$3\mathbf{a} + 5\mathbf{b} - 2\mathbf{c}$; $(-5/6, 1/3, -1/6)$]

9. **a**, **b**, **c** are non-coplanar base vectors. Show that the vectors $\mathbf{r}_1 = 3\mathbf{a} + \mathbf{b} - \mathbf{c}$, $\mathbf{r}_2 = -5\mathbf{a} + 2\mathbf{b} - 3\mathbf{c}$ and $\mathbf{r}_3 = 36\mathbf{a} + \mathbf{b} + 2\mathbf{c}$ are linearly dependent.
[$\mathbf{r}_3 = 7\mathbf{r}_1 - 3\mathbf{r}_2$]

10. Show that the vectors

(*a*) $\mathbf{a} = \mathbf{i} - 4\mathbf{k}$, $\mathbf{b} = 4\mathbf{i} + 3\mathbf{j} - \mathbf{k}$, $\mathbf{c} = 2\mathbf{i} + \mathbf{j} - 3\mathbf{k}$ are linearly dependent.
[$2\mathbf{a} + \mathbf{b} = 3\mathbf{c}$]

(*b*) $\mathbf{a} = 3\mathbf{i} + 2\mathbf{j} - \mathbf{k}$, $\mathbf{b} = \mathbf{i} - 3\mathbf{j} + \mathbf{k}$, $\mathbf{c} = 2\mathbf{i} + \mathbf{j} - 3\mathbf{k}$ are linearly independent.

11. Show that

$$\begin{vmatrix} x_1 & y_1 & z_1 \\ x_2 & y_2 & z_2 \\ x_3 & y_3 & z_3 \end{vmatrix} \neq 0$$

is a necessary and sufficient condition that the vectors $\mathbf{a} = x_1\mathbf{i} + y_1\mathbf{j} + z_1\mathbf{k}$, $\mathbf{b} = x_2\mathbf{i} + y_2\mathbf{j} + z_2\mathbf{k}$, $\mathbf{c} = x_3\mathbf{i} + y_3\mathbf{j} + z_3\mathbf{k}$ shall be linearly independent.

12. Show that the equation of the plane passing through the points A, B, C may be written

$$\mathbf{r} = (l\mathbf{a} + m\mathbf{b} + n\mathbf{c})/(l + m + n)$$

and verify that this equation is independent of the origin.

13. Show that the mid-points of the six edges of a cube which do not meet a given diagonal are coplanar.

14. The triangles ABC, A′B′C′ are such that AA′, BB′, CC′ are concurrent. AB and A′B′ meet at X and the other pairs of corresponding sides of the triangles meet at Y and Z. Show that X, Y, Z are collinear (Desargues' Theorem).

2 Scalar and vector products

2.1 Scalar Product of Two Vectors

The work done by a force **F** when its point of application is displaced from O to R (Fig. 2.1), i.e. is given a displacement **r**, is given by

$$\text{Work done} = F\cos\theta \times r$$

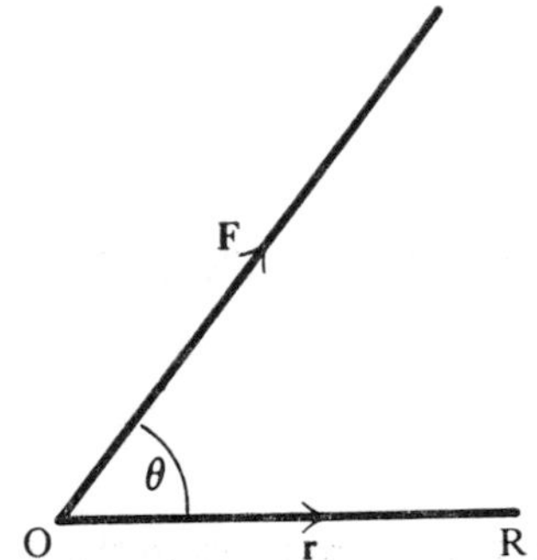

Figure 2.1

i.e. the product of the magnitudes of the two vectors **F** and **r** and the cosine of the angle between their directions.

This *scalar* quantity is said to be the scalar product of the vectors **F** and **r**.

DEFINITION 2.1
The **scalar product** of two vectors **A** and **B** is $AB\cos\theta$, where θ is the angle between the directions of **A** and **B**, and is denoted by **A.B** (Fig. 2.2).

Thus $\mathbf{A}.\mathbf{B} = AB \cos \theta = \mathbf{B}.\mathbf{A}$ $(0 \leqslant \theta \leqslant \pi)$.

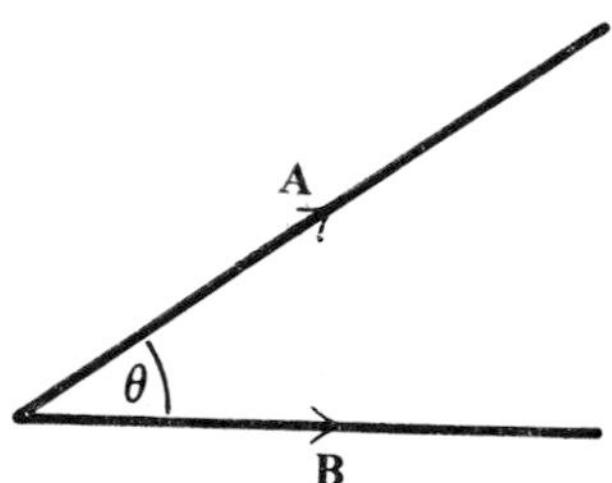

Figure 2.2

If **A** and **B** are perpendicular vectors, then $\mathbf{A}.\mathbf{B} = 0$. When $\mathbf{A} = \mathbf{B}$, we have $\mathbf{A}.\mathbf{A} = A^2$, the square of **A**.

The square of a unit vector is therefore 1.

$$\mathbf{i}.\mathbf{i} = i^2 = 1 \qquad \text{and} \qquad \mathbf{i}.\mathbf{i} = \mathbf{j}.\mathbf{j} = \mathbf{k}.\mathbf{k} = 1 \tag{2.1}$$

Since **i**, **j**, **k** are mutually perpendicular,

$$\mathbf{i}.\mathbf{j} = \mathbf{j}.\mathbf{k} = \mathbf{k}.\mathbf{i} = 0 \tag{2.2}$$

If m is a scalar,

$$m(\mathbf{A}.\mathbf{B}) = mAB \cos \theta = (m\mathbf{A}).\mathbf{B} = \mathbf{A}.m\mathbf{B}$$

Let **b** be a unit vector in the direction OB and let PQ be parallel to OB (Fig. 2.3). The projection of the vector **A** on OB is

$$\text{ON} = \text{PQ} = A \cos \theta = \mathbf{A}.\mathbf{b}$$

Let **r** be the vector from the origin O of rectangular coordinates to the point (x, y, z). Then

$$x = \text{the projection of } \mathbf{r} \text{ on } Ox = \mathbf{r}.\mathbf{i}$$

Similarly $y = \mathbf{r}.\mathbf{j}$ and $z = \mathbf{r}.\mathbf{k}$. Therefore

$$\mathbf{r} = x\mathbf{i} + y\mathbf{j} + z\mathbf{k} = (\mathbf{r}.\mathbf{i})\mathbf{i} + (\mathbf{r}.\mathbf{j})\mathbf{j} + (\mathbf{r}.\mathbf{k})\mathbf{k}$$

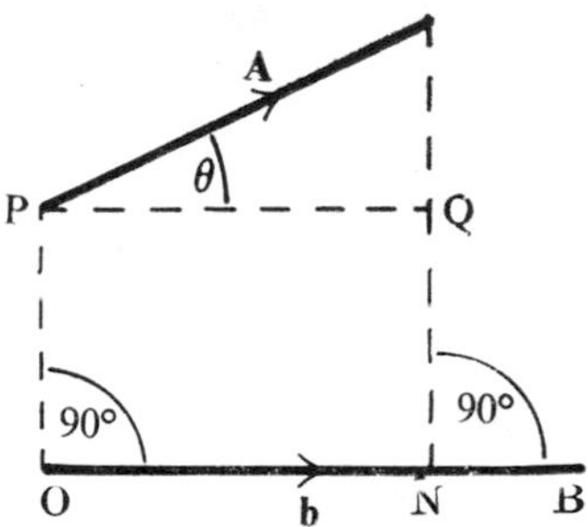

Figure 2.3

Note that the scalar products here are the scalar coefficients of the vectors **i**, **j**, **k**.

DISTRIBUTIVE LAW

With reference to Fig. 2.4, let **C** be a vector in the direction PQR and let **c** be a unit vector in the same direction. The projection of **A** + **B** in the direction PQR is equal to the sum of the projections of **A** and **B** in this direction. Therefore

$$(\mathbf{A} + \mathbf{B}).\mathbf{c} = \mathbf{A}.\mathbf{c} + \mathbf{B}.\mathbf{c}$$

On multiplying by C, we have

$$(\mathbf{A} + \mathbf{B}).C\mathbf{c} = \mathbf{A}.C\mathbf{c} + \mathbf{B}.C\mathbf{c}$$

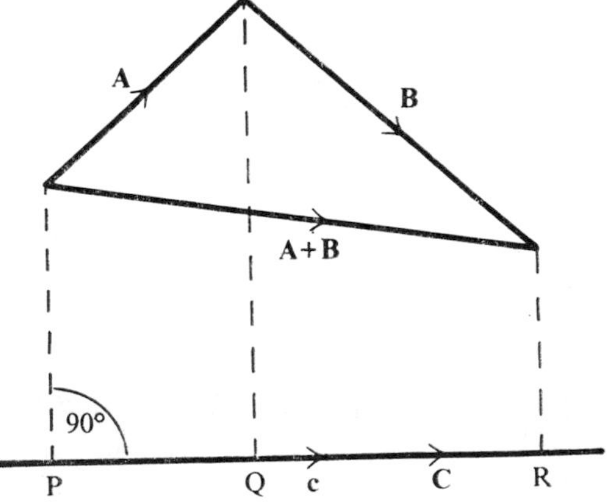

Figure 2.4

i.e.

$$(\mathbf{A} + \mathbf{B}).\mathbf{C} = \mathbf{A}.\mathbf{C} + \mathbf{B}.\mathbf{C} \tag{2.3}$$

so that the *distributive law* holds for scalar products.

By repeated applications of this result, it follows that the scalar produc of two sums of vectors may be expanded as in ordinary algebra, e.g.

$$\begin{aligned}(\mathbf{A} + \mathbf{B}).(\mathbf{C} + \mathbf{D}) &= (\mathbf{A} + \mathbf{B}).\mathbf{C} + (\mathbf{A} + \mathbf{B}).\mathbf{D} \\ &= \mathbf{A}.\mathbf{C} + \mathbf{B}.\mathbf{C} + \mathbf{A}.\mathbf{D} + \mathbf{B}.\mathbf{D}\end{aligned}$$

Let **A** and **B** be expressed in terms of their rectangular components, the

$$\begin{aligned}\mathbf{A}.\mathbf{B} &= (A_1\mathbf{i} + A_2\mathbf{j} + A_3\mathbf{k}).(B_1\mathbf{i} + B_2\mathbf{j} + B_3\mathbf{k}) \\ &= A_1B_1 + A_2B_2 + A_3B_3 \end{aligned} \tag{2.4}$$

Thus $A = \sqrt{(A_1^2 + A_2^2 + A_3^2)}$.

Since $\mathbf{A}.\mathbf{B} = AB\cos\theta$, it follows that the angle θ between the direction of **A** and **B** is given by

$$\begin{aligned}\cos\theta &= \frac{A_1B_1 + A_2B_2 + A_3B_3}{\sqrt{(A_1^2 + A_2^2 + A_3^2)}\sqrt{(B_1^2 + B_2^2 + B_3^2)}} \\ &= l_1l_2 + m_1m_2 + n_1n_2\end{aligned} \tag{2.5}$$

where (l_1, m_1, n_1), (l_2, m_2, n_2) are the direction cosines of the two vectors, i.e

$$l_1 = \cos\theta_1 = A_1/|\mathbf{A}| = A_1/\sqrt{(A_1^2 + A_2^2 + A_3^2)} \qquad \text{etc.}$$

EXAMPLE 2.1

(A) Find the angle between the vectors $\mathbf{A} = 3\mathbf{i} + \mathbf{j} - \mathbf{k}$ and $\mathbf{B} = 2\mathbf{i} - 3\mathbf{j} + 5\mathbf{k}$

$$\cos\theta = \frac{\mathbf{A}.\mathbf{B}}{AB} = \frac{3.2 + 1.(-3) + (-1).5}{\sqrt{[3^2 + 1^2 + (-1)^2]}\sqrt{[2^2 + (-3)^2 + 5^2]}} = \frac{-2}{\sqrt{11}\sqrt{38}}$$

$$\cos\theta = -0.0978 \qquad \theta = 95°37'$$

(B) What is the work done by a force $\mathbf{F} = 4\mathbf{i} + 3\mathbf{j} - \mathbf{k}$ in moving a particle along the vector $\mathbf{r} = 3\mathbf{i} + \mathbf{j} - 2\mathbf{k}$?

$$\text{Work done} = \mathbf{F}.\mathbf{r} = 4.3 + 3.1 + (-1)(-2) = 17$$

(C) Prove the formulae:
(i) $b = c \cos A + a \cos C$
(ii) $a^2 = b^2 + c^2 - 2bc \cos A$ for a triangle ABC (Fig. 2.5)

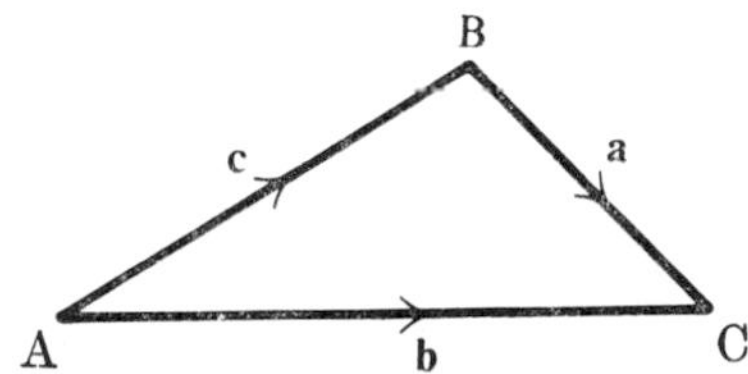

Figure 2.5

(i) $\mathbf{b} = \mathbf{c} + \mathbf{a}$ therefore $\mathbf{b.b} = \mathbf{c.b} + \mathbf{a.b}$
$b^2 = cb \cos A + ab \cos C \qquad b = c \cos A + a \cos C$
(ii) $\mathbf{a} = \mathbf{b} - \mathbf{c}$ therefore $\mathbf{a.a} = (\mathbf{b} - \mathbf{c}).(\mathbf{b} - \mathbf{c})$
$a^2 = b^2 + c^2 - 2\mathbf{b.c} = b^2 + c^2 - 2bc \cos A$

EXAMPLE 2.2

Define the projection of any point in three-dimensional Euclidean space onto a plane through the origin. If $\mathbf{u}$ is the position vector of the point and $p(\mathbf{u})$ denotes its projection on the plane, show that your definition can be expressed vectorially in the form

$$p(\mathbf{u}) = \mathbf{u} - (\mathbf{u.n})\mathbf{n}$$

where $\mathbf{n}$ is a unit vector perpendicular to the given plane.

Prove that $p^2 = p$.

Two such projections p_1, p_2 onto planes Π_1 and Π_2 through the origin,

$$p_1(\mathbf{u}) = \mathbf{u} - (\mathbf{u.n}_1)\mathbf{n}_1 \qquad p_2(\mathbf{u}) = \mathbf{u} - (\mathbf{u.n}_2)\mathbf{n}_2$$

are given. Prove that $p_1p_2 = p_2p_1$ if and only if Π_1 and Π_2 coincide or are perpendicular. (*Oxford and Cambridge: G.C.E. A-Level, S.M.P.*)

Solution Let R be a point whose position vector is $\mathbf{u}$. Let N be the point in which a perpendicular from R to the plane meets it (Fig. 2.6). Let $\mathbf{n}$ be a unit vector normal to the plane in the direction NR. Projection of **OR** on the plane is

$$\mathbf{ON} = \mathbf{OR} - \mathbf{NR} = \mathbf{u} - (\mathbf{u.n})\mathbf{n}$$

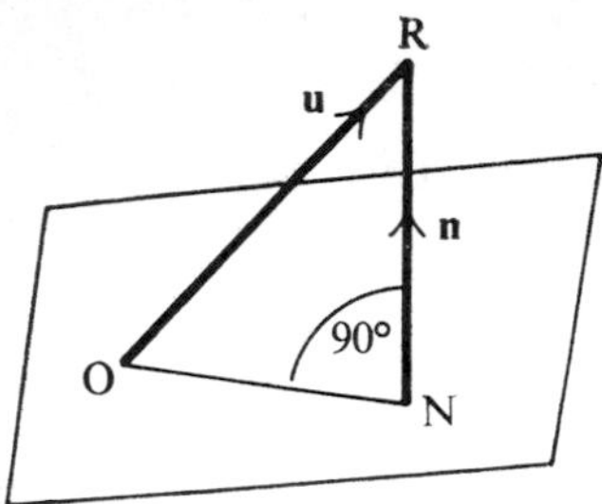

Figure 2.6

i.e. $p(\mathbf{u}) = \mathbf{u} - (\mathbf{u.n})\mathbf{n}$
Therefore

$$\begin{aligned} p^2(\mathbf{u}) = p\{p(\mathbf{u})\} &= \{\mathbf{u} - (\mathbf{u.n})\mathbf{n}\} - [\{\mathbf{u} - (\mathbf{u.n})\mathbf{n}\}.\mathbf{n}]\mathbf{n} \\ &= \mathbf{u} - (\mathbf{u.n})\mathbf{n} - [(\mathbf{u.n}) - (\mathbf{u.n})(\mathbf{n.n})]\mathbf{n} \\ &= \mathbf{u} - (\mathbf{u.n})\mathbf{n} \end{aligned}$$

i.e. $p^2(\mathbf{u}) = p(\mathbf{u})$. (It is otherwise obvious that the projection of **ON** on the plane is **ON**.) Now,

$$\begin{aligned} p_1p_2(\mathbf{u}) = p_1\{p_2(\mathbf{u})\} &= \mathbf{u} - (\mathbf{u.n}_2)\mathbf{n}_2 - [\{\mathbf{u} - (\mathbf{u.n}_2)\mathbf{n}_2\}.\mathbf{n}_1]\mathbf{n}_1 \\ &= \mathbf{u} - (\mathbf{u.n}_2)\mathbf{n}_2 - (\mathbf{u.n}_1)\mathbf{n}_1 + (\mathbf{u.n}_2)(\mathbf{n}_2.\mathbf{n}_1)\mathbf{n}_1 \end{aligned}$$

Similarly $\quad p_2p_1(\mathbf{u}) = \mathbf{u} - (\mathbf{u.n}_1)\mathbf{n}_1 - (\mathbf{u.n}_2)\mathbf{n}_2 + (\mathbf{u.n}_1)(\mathbf{n}_1.\mathbf{n}_2)\mathbf{n}_2$

Hence, $p_1p_2 = p_2p_1$ provided that $(\mathbf{u.n}_2)(\mathbf{n}_1.\mathbf{n}_2)\mathbf{n}_1 = (\mathbf{u.n}_1)(\mathbf{n}_1.\mathbf{n}_2)\mathbf{n}_2$
i.e. if $\mathbf{n}_1 = \mathbf{n}_2$ when the planes coincide
or if $\mathbf{n}_1.\mathbf{n}_2 = 0$ when the planes are perpendicular.

EXAMPLE 2.3
Prove that the perpendiculars drawn from the vertices of a triangle to the opposite sides are concurrent.

Solution Let the perpendiculars from A and B to the opposite sides meet in H (Fig. 2.7).

Since AH is perpendicular to BC, we have $(\mathbf{h} - \mathbf{a}).(\mathbf{b} - \mathbf{c}) = 0$.

$$\mathbf{h.b} - \mathbf{a.b} - \mathbf{h.c} + \mathbf{a.c} = 0 \qquad \text{(i)}$$

Similarly $(\mathbf{h} - \mathbf{b}).(\mathbf{c} - \mathbf{a}) = 0$

$$\mathbf{h.c} - \mathbf{b.c} - \mathbf{h.a} + \mathbf{a.b} = 0 \qquad \text{(ii)}$$

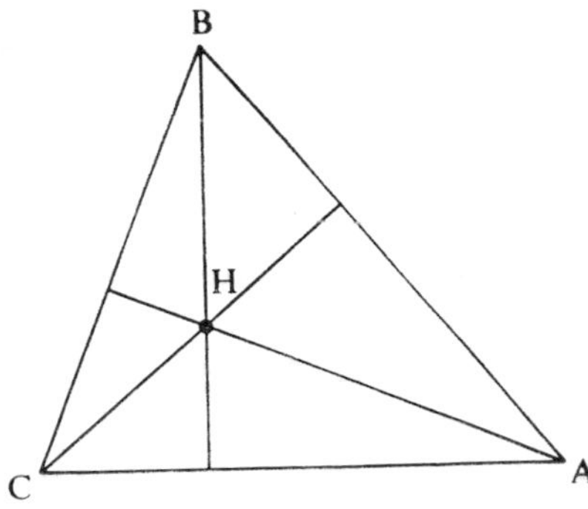

Figure 2.7

On adding equations (i) and (ii), we have

$$(\mathbf{h} - \mathbf{c}).(\mathbf{b} - \mathbf{a}) = 0$$

so that HC is perpendicular to AB which proves the theorem.

EXAMPLE 2.4

Two pairs of opposite edges of a tetrahedron are perpendicular. Show that the third pair are also perpendicular. Show also that the sum of the squares of the lengths of the edges is equal to four times the sum of the squares of the lengths of the lines joining the mid-points of the opposite edges.

Solution Let OA be perpendicular to BC (Fig. 2.8). Then

$$\mathbf{a}.(\mathbf{c} - \mathbf{b}) = 0 \qquad \mathbf{a}.\mathbf{c} = \mathbf{a}.\mathbf{b} \qquad \text{(i)}$$

Let OB be perpendicular to AC. Then

$$\mathbf{b}.(\mathbf{c} - \mathbf{a}) = 0 \qquad \mathbf{b}.\mathbf{c} = \mathbf{a}.\mathbf{b} \qquad \text{(ii)}$$

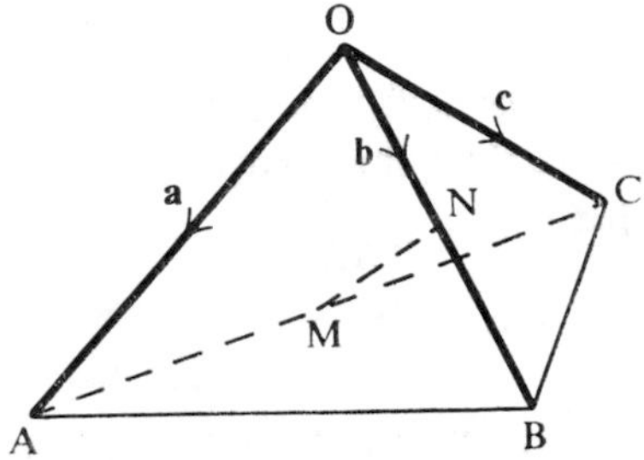

Figure 2.8

Therefore

$$\mathbf{a.c} = \mathbf{b.c} \qquad \mathbf{c.(b - a)} = 0 \qquad \text{(iii)}$$

Therefore OC is perpendicular to AB.

Let M and N be the mid-points of AC and OB respectively. Then

$$\mathbf{MN} = \mathbf{ON} - \mathbf{OM} = \tfrac{1}{2}\mathbf{b} - \tfrac{1}{2}(\mathbf{a} + \mathbf{c}) = \tfrac{1}{2}\{(\mathbf{b} - \mathbf{c}) - \mathbf{a}\}$$

Sum of the squares of the joins of the mid-points of opposite edges is

$$\begin{aligned}
&\tfrac{1}{4}[\{(\mathbf{b} - \mathbf{c}) - \mathbf{a}\}^2 + \{(\mathbf{c} - \mathbf{a}) - \mathbf{b}\}^2 + \{(\mathbf{a} - \mathbf{b}) - \mathbf{c}\}^2] \\
&= \tfrac{1}{4}[(\mathbf{b} - \mathbf{c})^2 + (\mathbf{c} - \mathbf{a})^2 + (\mathbf{a} - \mathbf{b})^2 + \mathbf{a}^2 + \mathbf{b}^2 + \mathbf{c}^2 \\
&\qquad - 2\{\mathbf{a.(b - c)} + \mathbf{b.(c - a)} + \mathbf{c.(a - b)}\}] \\
&= \tfrac{1}{4}[\mathbf{a}^2 + \mathbf{b}^2 + \mathbf{c}^2 + (\mathbf{b} - \mathbf{c})^2 + (\mathbf{c} - \mathbf{a})^2 + (\mathbf{a} - \mathbf{b})^2] \\
&\qquad\qquad \text{[by (i), (ii) and (iii)]} \\
&= \tfrac{1}{4}[\text{sum of the squares of the lengths of the edges}]
\end{aligned}$$

EXERCISES 2.1

1. Calculate the work done by a force of 30 newtons whose line of action has direction cosines $(\frac{2}{3}, \frac{1}{3}, \frac{2}{3})$ in a displacement from the point (1, 3, 5) to the point (7, 9, 2) where distances are measured in metres. [120 joules]
2. What is the projection of the vector $\mathbf{A} = 2\mathbf{i} - 3\mathbf{j} + \mathbf{k}$ on the vector $\mathbf{B} = 4\mathbf{i} - 7\mathbf{j} + 4\mathbf{k}$? [11/3]
3. If **a**, **b**, **c** are coplanar vectors and **a** is not parallel to **b**, show that

$$\mathbf{c} = \frac{\begin{vmatrix} \mathbf{c.a} & \mathbf{a.b} \\ \mathbf{c.b} & \mathbf{b.b} \end{vmatrix} \mathbf{a} + \begin{vmatrix} \mathbf{a.a} & \mathbf{c.a} \\ \mathbf{a.b} & \mathbf{c.b} \end{vmatrix} \mathbf{b}}{\begin{vmatrix} \mathbf{a.a} & \mathbf{a.b} \\ \mathbf{a.b} & \mathbf{b.b} \end{vmatrix}}$$

4. Prove that the diagonals of a rhombus intersect at right angles.
5. Show that the perpendicular bisectors of the sides of a triangle are concurrent.
6. For the tetrahedron in Fig. 2.8, show that

$$\mathrm{OA}^2 + \mathrm{BC}^2 = \mathrm{OB}^2 + \mathrm{CA}^2 = \mathrm{OC}^2 + \mathrm{AB}^2$$

7. Show that, in a regular tetrahedron, the perpendiculars from the vertices to the opposite faces meet these faces in their centroids. Show that the angle between

two faces is $\cos^{-1} 1/3$ and that the angle between a face and an edge not in that face is $\cos^{-1} 1/\sqrt{3}$.

8. A tetrahedron OABC has a vertex O at the origin and adjacent edges OA, OB, OC are represented by the vectors **a**, **b**, **c** respectively. If G is the centroid of the face ABC, prove that

$$3\mathbf{OG} = \mathbf{a} + \mathbf{b} + \mathbf{c}$$

If the angles BOC, COA, AOB are α, β, γ respectively and if the lengths of OA, OB, OC are a, b, c respectively, prove that

$$9\mathrm{OG}^2 = a^2 + b^2 + c^2 + 2bc \cos \alpha + 2ac \cos \beta + 2ab \cos \gamma$$

Find also an expression for the cosine of the angle between **AB** and **OC** (*U.L.A-level*).

$[(a \cos \beta - b \cos \alpha)(a^2 + b^2 - 2ab \cos \gamma)^{-1/2}]$

9. Define the scalar product of two three-dimensional Euclidean vectors **u**, **v**. Deduce an expression for the angle between the vectors

$$\mathbf{u} = u_1\mathbf{i} + u_2\mathbf{j} + u_3\mathbf{k} \quad \text{and} \quad \mathbf{v} = v_1\mathbf{i} + v_2\mathbf{j} + v_3\mathbf{k}$$

in terms of u_1, u_2, u_3, v_1, v_2, v_3. A regular tetrahedron has vertices O, A, B, C where O is the origin, and A, B, C have position vectors with respect to O given by

$$\mathbf{OA} = -\mathbf{i} + \mathbf{j} \qquad \mathbf{OB} = a\mathbf{i} + b\mathbf{j} \qquad \mathbf{OC} = p\mathbf{i} + q\mathbf{j} + r\mathbf{k}$$

Find numerical values of a, b, p, q, r given that $a > 0$ and $r > 0$. (*Oxford and Cambridge:* G.C.E. A-Level, S.M.P)

$[a = \frac{1}{2}(\sqrt{3} - 1); b = \frac{1}{2}(\sqrt{3} + 1); p = -\frac{1}{2} + \sqrt{3}/6; q = \frac{1}{2} + \sqrt{3}/6; r = \frac{1}{3}\sqrt{3}]$

2.2 *Equation of a Plane*

Let the plane pass through a point A and let **n** be a unit vector perpendicular to the plane and having the direction from the origin O *towards* the plane (Fig. 2.9). Let R be any point on the plane so that $\mathbf{AR} = \mathbf{r} - \mathbf{a}$ is perpendicular to **n**.

Therefore $(\mathbf{r} - \mathbf{a}).\mathbf{n} = 0$.

Since **n** has the direction of **ON**, it follows that $\mathbf{a}.\mathbf{n} = \mathrm{ON} = p$, a positive number, which is the length of the perpendicular from the origin to the plane.

Therefore the equation of the plane takes the form

$$\mathbf{r}.\mathbf{n} = p \text{ (normal form)} \tag{2.6}$$

or

$$lx + my + nz = p \tag{2.7}$$

in Cartesian coordinates where l, m, n are the direction cosines of the normal ON.

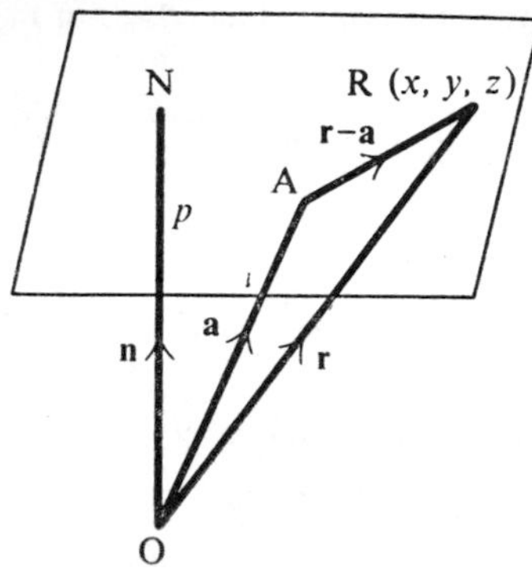

Figure 2.9

The angle of inclination of two planes is the angle θ between their normals. Therefore

$$\cos\theta = \mathbf{n}_1.\mathbf{n}_2 = l_1l_2 + m_1m_2 + n_1n_2 \tag{2.8}$$

2.3 *Perpendicular Distance of a Point from a Plane*

Let NM (Fig. 2.10) be the trace of the plane $\mathbf{r}.\mathbf{n} = p$. Suppose we require the perpendicular distance $R'S = d$ of the point R′ from the plane NM.

Let N′M′ be a plane through R′ parallel to NM. The equation of N′M′ is

$$\mathbf{r}.\mathbf{n} = p' \quad \text{where} \quad p' = ON'$$

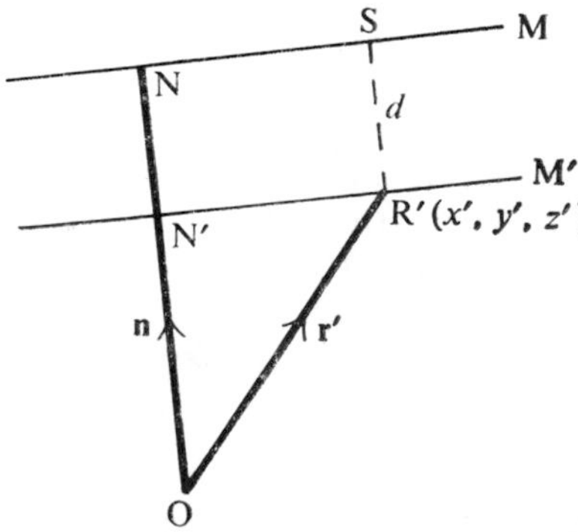

Figure 2.10

Therefore $\mathbf{r}'.\mathbf{n} = p'$ and

$$d = \mathrm{R'S} = \mathrm{N'N} = p - p' = p - \mathbf{r}'.\mathbf{n} \tag{2.9}$$

This quantity will be positive for points R′ on the origin side of the plane $\mathbf{r}.\mathbf{n} = p$ and negative for points on the other side.

In Cartesian coordinates

$$d = p - (lx' + my' + nz') \tag{2.10}$$

2.4 *The Equations of Planes Bisecting the Angles between Two Given Planes*

Let the two given planes be $\mathbf{r}.\mathbf{n}_1 = p_1$ and $\mathbf{r}.\mathbf{n}_2 = p_2$.

Any point on a bisector will be equidistant from these two planes. The perpendicular distances of any point on the bisector of the angle containing the origin will both be positive. Therefore the equation of this bisecting plane will be (eqn (2.9))

$$p_1 - \mathbf{r}.\mathbf{n}_1 = p_2 - \mathbf{r}.\mathbf{n}_2$$

i.e.

$$\mathbf{r}.(\mathbf{n}_1 - \mathbf{n}_2) = p_1 - p_2 \tag{2.11}$$

For any point on the other bisecting plane, the perpendicular distances will have opposite signs, so that its equation will be

$$\mathbf{r}.(\mathbf{n}_1 + \mathbf{n}_2) = p_1 + p_2 \tag{2.12}$$

2.5 *Vector Area*

A plane area may be represented by a vector. Let A be the magnitude of the area. Let $\mathbf{n}$ be a unit vector normal to this area. To distinguish between the two possible directions of $\mathbf{n}$, the following convention is adopted. Let $\mathbf{n}$ have the direction of advance of a right-handed screw which is rotated in the direction PQR in which the boundary of the area is described (Fig. 2.11). This means that $\mathbf{n}$ will have the direction of OZ in a right-handed system of coordinate axes when the boundary of the area is described in a direction from Ox towards Oy in the first quadrant.

The area A may then be represented by the vector area $A\mathbf{n} = \mathbf{A}$.

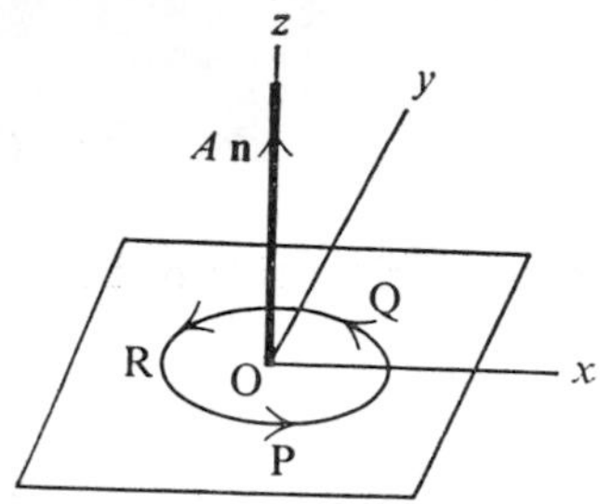

Figure 2.11

2.6 Vector Rotation

A rotation may be represented by a vector whose direction is that of the axis of rotation and whose magnitude is that of the angle of rotation, say θ. The direction of this vector is that of a unit vector $\mathbf{n}$ as determined by the convention of Section 2.5 where the direction of the rotation is from Ox towards Oy in the first quadrant (Fig. 2.12).

Thus the rotation θ may be represented by the vector rotation $\theta\mathbf{n} = \boldsymbol{\theta}$.

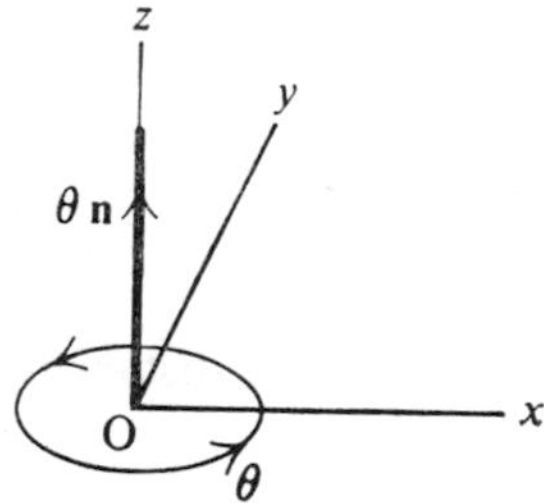

Figure 2.12

EXAMPLE 2.5

Find the equation of the plane through the point A (2, −1, 3) perpendicular to the line OB where B is the point (3, −2, −6). What is the length of the perpendicular from the origin to this plane? What is the perpendicular distance from the point (4, 3, −2) to the plane?

Solution

$$\mathbf{a} = 2\mathbf{i} - \mathbf{j} + 3\mathbf{k} \qquad \mathbf{b} = 3\mathbf{i} - 2\mathbf{j} - 6\mathbf{k}$$

Equation of the plane is $(\mathbf{r} - \mathbf{a}).\mathbf{b} = 0$, i.e. $\mathbf{r}.\mathbf{b} = \mathbf{a}.\mathbf{b}$. Substituting

$$(x\mathbf{i} + y\mathbf{j} + z\mathbf{k}).(3\mathbf{i} - 2\mathbf{j} - 6\mathbf{k}) = (2\mathbf{i} - \mathbf{j} + 3\mathbf{k}).(3\mathbf{i} - 2\mathbf{j} - 6\mathbf{k})$$

$$3x - 2y - 6z = 6 + 2 - 18 = -10$$

$$3x - 2y - 6z + 10 = 0$$

This equation will be converted to the normal form (see eqn (2.7)) upon division by $\sqrt{\{(3^2) + (-2)^2 + (-6)^2\}}$, i.e. by 7.

Normal form is $-\frac{3}{7}x + \frac{2}{7}y + \frac{6}{7}z = \frac{10}{7}$. Thus the perpendicular distance from the origin to the plane is $p = 10/7$.

Alternatively: $\mathbf{n} = (-3\mathbf{i} + 2\mathbf{j} + 6\mathbf{k})/7$

$$p = \mathbf{a}.\mathbf{n} = [(2\mathbf{i} - \mathbf{j} + 3\mathbf{k}).(-3\mathbf{i} + 2\mathbf{j} + 6\mathbf{k})]/7 = 10/7$$

Distance d of the point R′ $[\mathbf{r}' = 4\mathbf{i} + 3\mathbf{j} - 2\mathbf{k}]$ from the plane is given by

$$d = p - \mathbf{r}'.\mathbf{n} = \frac{10}{7} - \frac{(4\mathbf{i} + 3\mathbf{j} - 2\mathbf{k}).(-3\mathbf{i} + 2\mathbf{j} + 6\mathbf{k})}{7} = \frac{10}{7} + \frac{18}{7} = 4$$

EXAMPLE 2.6

Find the dihedral angle between the two planes

$$2x + 6y - 3z = 10 \qquad 7x + 4y - 4z = 8$$

Find also the equation of the plane which bisects the dihedral angle containing the origin.

Solution The equations of the planes will be converted to the normal form on division by 7 and by 9 respectively.

$$\mathbf{n}_1 = [2\mathbf{i} + 6\mathbf{j} - 3\mathbf{k}]/7 \qquad \mathbf{n}_2 = [7\mathbf{i} + 4\mathbf{j} - 4\mathbf{k}]/9$$

$$\mathbf{n}_1 - \mathbf{n}_2 = \tfrac{1}{63}(-31\mathbf{i} + 26\mathbf{j} + \mathbf{k})$$

Also $p_1 = 10/7$ and $p_2 = 8/9$. Therefore the equation of the plane bisecting the dihedral angle containing the origin is given by eqn (2.11):

$$\mathbf{r}.(\mathbf{n}_1 - \mathbf{n}_2) = p_1 - p_2$$

i.e. $(x\mathbf{i} + y\mathbf{j} + z\mathbf{k}).(-31\mathbf{i} + 26\mathbf{j} + \mathbf{k})/63 = (10/7) - (8/9)$

$$-31x + 26y + z = 34$$

The dihedral angle θ is given by eqn (2.8):

$$\cos\theta = \mathbf{n}_1.\mathbf{n}_2 = \frac{2}{7}\cdot\frac{7}{9} + \frac{6}{7}\cdot\frac{4}{9} + \left(\frac{-3}{7}\right)\cdot\left(\frac{-4}{9}\right) = \frac{50}{63}$$

EXERCISES 2.2

1. Find the perpendicular distance of the point $(-1, 2, 3)$ from the plane $2x - 9y + 6z = 12$ [14/11]

2. Find the equation of a plane which passes through the point A (3, -1, 2) and is perpendicular to AB where B is the point (-5, 3, -1). What are the perpendicular distances from the origin and from the point (2, -3, 5) to the plane? [$8x - 4y + 3z = 34$; $34/\sqrt{89}$; $(-)\,9/\sqrt{89}$]
3. Find the equation of the plane passing through the origin and through the line of intersection of the planes

$$\mathbf{r.a} = \lambda \qquad \mathbf{r.b} = \mu$$

[$\mathbf{r}.(\mu\mathbf{a} - \lambda\mathbf{b}) = 0$]
4. Show that the equation of a sphere centre C and radius a is

$$\mathbf{r}^2 - 2\mathbf{r.c} + c^2 - a^2 = 0$$

where **c** is the position vector of C.
5. Find the equation of a sphere on AB as diameter.

[$(\mathbf{r} - \mathbf{a}).(\mathbf{r} - \mathbf{b}) = 0$]
6. If **c** is the position vector of the point (x_0, y_0, z_0), obtain the equations of
   (i) a sphere, centre (x_0, y_0, z_0), of radius a,
   (ii) a plane perpendicular to **c** passing through (x_0, y_0, z_0),
   (iii) a sphere, centre $(\frac{1}{2}x_0, \frac{1}{2}y_0, \frac{1}{2}z_0)$, passing through the origin.
   [(i) $|\mathbf{r} - \mathbf{c}| = a$, (ii) $(\mathbf{r} - \mathbf{c}).\mathbf{c} = 0$, (iii) $(\mathbf{r} - \mathbf{c}).\mathbf{r} = 0$]

2.7 *Vector Product of Two Vectors*

Let **F** be a force localized in the line NR and let **r** be the position vector of the point R on this line with respect to a point O (Fig. 2.13).

The moment of **F** about O has a magnitude $F \times \mathrm{ON} = Fr \sin\theta$.

Let **n** be a unit vector normal to the plane of **r** and **F** such that **r**, **F** and **n** form a right-handed system (Fig. 2.14).

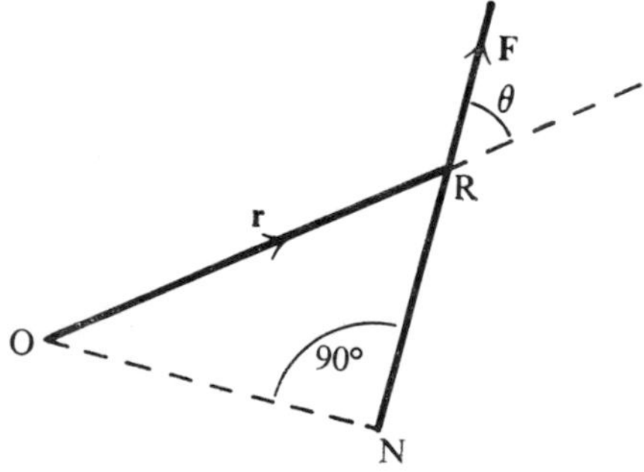

Figure 2.13

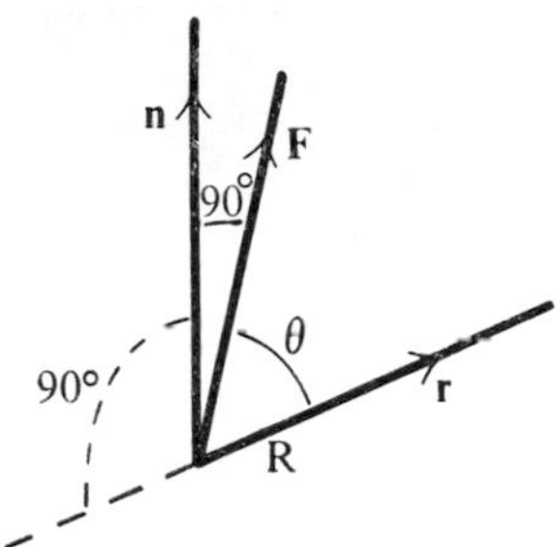

Figure 2.14

It is convenient to represent the moment or torque of the force **F** about O by the vector

$$\mathbf{G} = (Fr \sin \theta)\mathbf{n}$$

which is called the vector product of **r** and **F** and is denoted by $\mathbf{r} \times \mathbf{F}$, i.e.

$$\mathbf{G} = \mathbf{r} \times \mathbf{F} = (Fr \sin \theta)\mathbf{n}$$

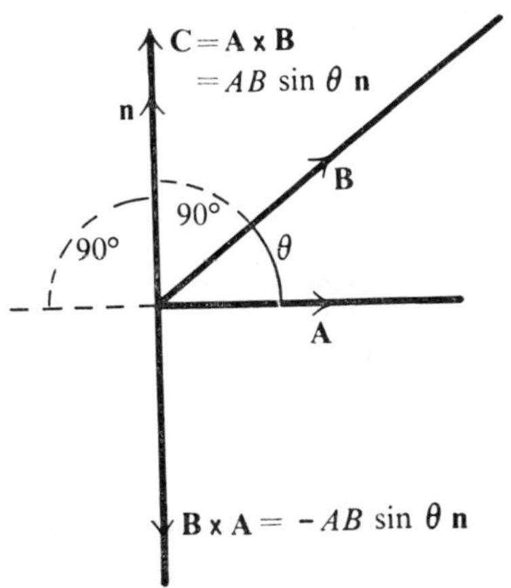

Figure 2.15

Definition 11.2
The vector product of two vectors **A** and **B** is a vector $\mathbf{C} = \mathbf{A} \times \mathbf{B}$ (Fig. 2.15).

The magnitude of **C** is defined to be $AB \sin \theta$ where θ is the angle between the directions of the vectors **A** and **B**. The direction of **C** is perpendicular to the plane of **A** and **B** such that **A**, **B** and **C** form a right-handed system.

Therefore

$$\mathbf{C} = \mathbf{A} \times \mathbf{B} = AB \sin \theta \mathbf{n} \qquad (0 \leqslant \theta \leqslant \pi) \tag{2.13}$$

where $\mathbf{n}$ is a unit vector in the direction of $\mathbf{A} \times \mathbf{B}$.

It follows that $\mathbf{B} \times \mathbf{A}$ has the opposite direction to $\mathbf{A} \times \mathbf{B}$ but has the same magnitude. Therefore

$$\mathbf{B} \times \mathbf{A} = -\mathbf{A} \times \mathbf{B}$$

Thus the Commutative Law does not hold for vector products.

If $\mathbf{A}$ and $\mathbf{B}$ are parallel vectors, $\sin \theta = 0$. Therefore

$$\mathbf{A} \times \mathbf{B} = 0$$

In particular, $\mathbf{A} \times \mathbf{A} = 0$ for all vectors $\mathbf{A}$.

It follows that for the unit vectors $\mathbf{i}, \mathbf{j}, \mathbf{k}$,

$$\mathbf{i} \times \mathbf{i} = \mathbf{j} \times \mathbf{j} = \mathbf{k} \times \mathbf{k} = 0 \tag{2.14}$$

whilst

$$\begin{aligned} \mathbf{i} \times \mathbf{j} &= -\mathbf{j} \times \mathbf{i} = \mathbf{k} \\ \mathbf{j} \times \mathbf{k} &= -\mathbf{k} \times \mathbf{j} = \mathbf{i} \\ \mathbf{k} \times \mathbf{i} &= -\mathbf{i} \times \mathbf{k} = \mathbf{j} \end{aligned} \tag{2.15}$$

For any scalar m,

$$m\mathbf{A} \times \mathbf{B} = m\mathbf{AB} \sin \theta \mathbf{n} = \mathbf{A} \times m\mathbf{B} = m(\mathbf{A} \times \mathbf{B})$$

DISTRIBUTIVE LAW

Let $\mathbf{A}$ be a vector which is perpendicular to each of two vectors $\mathbf{B}_1$ and $\mathbf{C}_1$.

To prove that

$$\mathbf{A} \times (\mathbf{B}_1 + \mathbf{C}_1) = \mathbf{A} \times \mathbf{B}_1 + \mathbf{A} \times \mathbf{C}_1$$

The vector $\mathbf{A} \times \mathbf{B}_1$ lies in the plane defined by $\mathbf{B}_1$ and $\mathbf{C}_1$, is perpendicular to $\mathbf{A}$ and $\mathbf{B}_1$, and has a magnitude AB_1 (Fig. 2.16).

The vectors $\mathbf{A} \times \mathbf{C}_1$ and $\mathbf{A} \times (\mathbf{B}_1 + \mathbf{C}_1)$ also lie in the plane defined by $\mathbf{B}_1$ and $\mathbf{C}_1$, are respectively perpendicular to $\mathbf{C}_1$ and $(\mathbf{B}_1 + \mathbf{C}_1)$, and have magnitudes AC_1 and $A\,|\mathbf{B}_1 + \mathbf{C}_1|$ respectively.

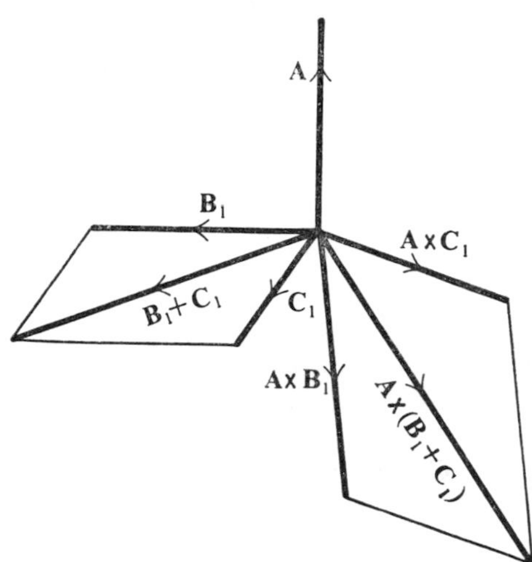

Figure 2.16

It follows that if the parallelogram whose sides are $\mathbf{B}_1$ and $\mathbf{C}_1$ and whose diagonal is $(\mathbf{B}_1 + \mathbf{C}_1)$ is rotated through $+90°$ about $\mathbf{A}$ and is then magnified A times, the resulting parallelogram has sides representing $\mathbf{A} \times \mathbf{B}_1$ and $\mathbf{A} \times \mathbf{C}_1$ and a diagonal representing $\mathbf{A} \times (\mathbf{B}_1 + \mathbf{C}_1)$. Therefore

$$\mathbf{A} \times (\mathbf{B}_1 + \mathbf{C}_1) = \mathbf{A} \times \mathbf{B}_1 + \mathbf{A} \times \mathbf{C}_1 \tag{2.16}$$

Now, to generalize, let $\mathbf{A}$, $\mathbf{B}$, $\mathbf{C}$ be non-coplanar vectors, and let $\mathbf{B}_1$, $\mathbf{B}_2$ be the components of $\mathbf{B}$, perpendicular and parallel respectively to $\mathbf{A}$ (Fig. 2.17).

The magnitude of $\mathbf{A} \times \mathbf{B}_1 = AB \sin\theta =$ magnitude of $\mathbf{A} \times \mathbf{B}$.

Also, the directions of $\mathbf{A} \times \mathbf{B}_1$ and $\mathbf{A} \times \mathbf{B}$ are the same. Therefore

$$\mathbf{A} \times \mathbf{B}_1 = \mathbf{A} \times \mathbf{B}$$

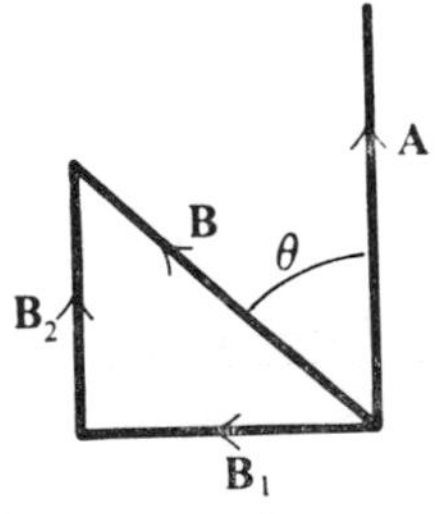

Figure 2.17

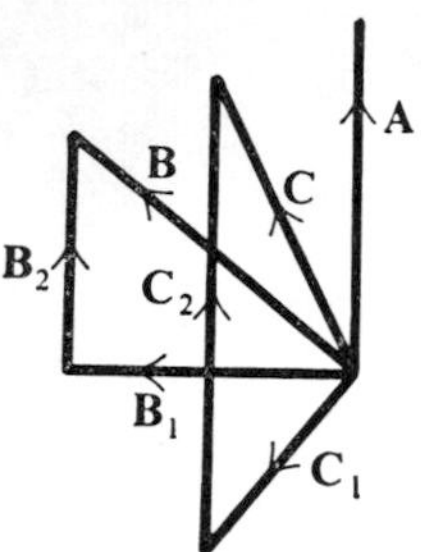

Figure 2.18

Similarly $\mathbf{A} \times \mathbf{C}_1 = \mathbf{A} \times \mathbf{C}$.

Also $\mathbf{B} + \mathbf{C} = (\mathbf{B}_1 + \mathbf{C}_1) + (\mathbf{B}_2 + \mathbf{C}_2)$. Therefore

$$\mathbf{A} \times (\mathbf{B} + \mathbf{C}) = \mathbf{A} \times (\mathbf{B}_1 + \mathbf{C}_1)$$

Now by eqn (2.16) it follows that

$$\mathbf{A} \times (\mathbf{B} + \mathbf{C}) = \mathbf{A} \times \mathbf{B} + \mathbf{A} \times \mathbf{C} \tag{2.17}$$

Expressing the vectors **A** and **B** in terms of their rectangular components and applying the distributive law, we have

$$\begin{aligned} \mathbf{A} \times \mathbf{B} &= (A_1\mathbf{i} + A_2\mathbf{j} + A_3\mathbf{k}) \times (B_1\mathbf{i} + B_2\mathbf{j} + B_3\mathbf{k}) \\ &= (A_2B_3 - A_3B_2)\mathbf{i} + (A_3B_1 - A_1B_3)\mathbf{j} + (A_1B_2 - A_2B_1)\mathbf{k} \end{aligned} \tag{2.18}$$

by virtue of eqns (2.14) and (2.15).

In determinantal form, this becomes

$$\mathbf{A} \times \mathbf{B} = \begin{vmatrix} \mathbf{i} & \mathbf{j} & \mathbf{k} \\ A_1 & A_2 & A_3 \\ B_1 & B_2 & B_3 \end{vmatrix} \tag{2.19}$$

Since $\mathbf{A} \times \mathbf{B} = AB \sin \theta \mathbf{n}$, it follows that

$$\sin^2 \theta = \frac{(A_2B_3 - A_3B_2)^2 + (A_3B_1 - A_1B_3)^2 + (A_1B_2 - A_2B_1)^2}{(A_1^2 + A_2^2 + A_3^2)(B_1^2 + B_2^2 + B_3^2)} \tag{2.20}$$

$$= (m_1n_2 - m_2n_1)^2 + (n_1l_2 - n_2l_1)^2 + (l_1m_2 - l_2m_1)^2 \tag{2.21}$$

where (l_1, m_1, n_1) and (l_2, m_2, n_2) are the direction cosines of **A** and **B** respectively.

EXAMPLE 2.7

(A) Find the vector of magnitude 10 which is perpendicular to each of the vectors $\mathbf{A} = 2\mathbf{i} - \mathbf{j} + \mathbf{k}$ and $\mathbf{B} = \mathbf{i} + 3\mathbf{k}$.

Solution The required vector will have the direction of

$$\mathbf{C} = \mathbf{A} \times \mathbf{B} = \begin{vmatrix} \mathbf{i} & \mathbf{j} & \mathbf{k} \\ 2 & -1 & 1 \\ 1 & 0 & 3 \end{vmatrix} = -3\mathbf{i} - 5\mathbf{j} + \mathbf{k}$$

A unit vector having the direction of **C** is $\dfrac{1}{\sqrt{35}}(-3\mathbf{i} - 5\mathbf{j} + \mathbf{k})$.

$$\text{Required vector is } \pm\frac{10}{\sqrt{35}}(3\mathbf{i} + 5\mathbf{j} - \mathbf{k}) = \pm\frac{2\sqrt{35}}{7}(3\mathbf{i} + 5\mathbf{j} - \mathbf{k}).$$

(B) Show that the area of a parallelogram having sides **A** and **B** is $|\mathbf{A} \times \mathbf{B}|$.

Solution Let h be the height of the parallelogram (Fig. 2.19).

$$\text{Area} = |\mathbf{A}|\,h = |\mathbf{B}| \sin\theta\, |\mathbf{A}| = |\mathbf{A} \times \mathbf{B}|$$

Thus the area of a triangle with sides **A** and **B** is $= \frac{1}{2}|\mathbf{A} \times \mathbf{B}|$.

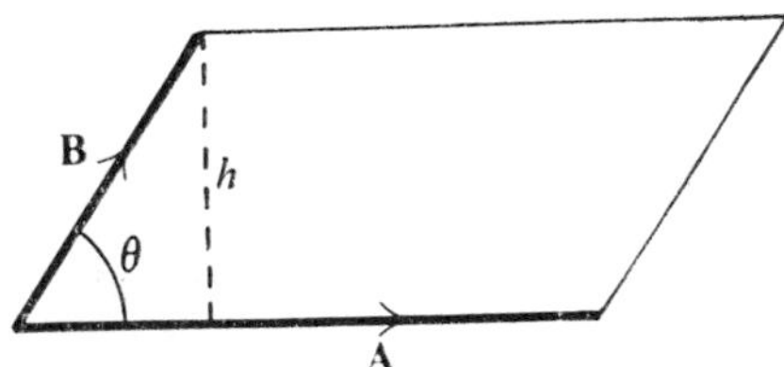

Figure 2.19

(C) Prove the sine law for a plane triangle.

Solution Let **a, b, c** be vectors representing the sides of the triangle ABC (Fig. 2.20). Therefore

$$\mathbf{a} + \mathbf{b} + \mathbf{c} = 0$$

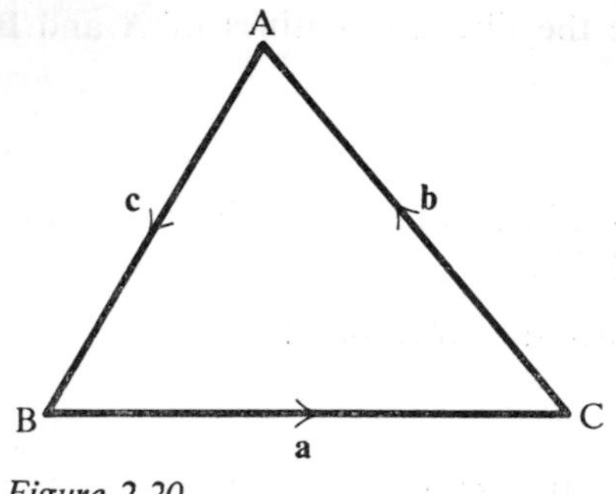

Figure 2.20

Take the vector product of each term by **a**,

$$\mathbf{a} \times \mathbf{a} + \mathbf{a} \times \mathbf{b} + \mathbf{a} \times \mathbf{c} = 0$$

i.e.

$$\mathbf{a} \times \mathbf{b} = \mathbf{c} \times \mathbf{a}$$

Similarly by taking the vector product by **b**, $\mathbf{a} \times \mathbf{b} = \mathbf{b} \times \mathbf{c}$. Therefore

$$\mathbf{a} \times \mathbf{b} = \mathbf{b} \times \mathbf{c} = \mathbf{c} \times \mathbf{a}$$

$$ab \sin C = bc \sin A = ca \sin B$$

$$\frac{\sin A}{a} = \frac{\sin B}{b} = \frac{\sin C}{c}$$

(This result also follows immediately from (B).)

(D) Show that the vector sum of the vector areas of the faces of a tetrahedron is zero.

Solution By (B), the area of the face OAB $= \frac{1}{2} |\mathbf{a} \times \mathbf{b}|$.

Thus vector area of the face OAB $= \frac{1}{2}(\mathbf{a} \times \mathbf{b})$ which has the direction of the *outward* normal to this face.

Similarly the vector areas of the other faces in the directions of their outward normals are

$$\tfrac{1}{2}(\mathbf{b} \times \mathbf{c}) \qquad \tfrac{1}{2}(\mathbf{c} \times \mathbf{a}) \qquad \tfrac{1}{2}\{(\mathbf{c} - \mathbf{a}) \times (\mathbf{b} - \mathbf{a})\}$$

Therefore, the sum of vector areas is

$$\begin{aligned} &\tfrac{1}{2}[\mathbf{a} \times \mathbf{b} + \mathbf{b} \times \mathbf{c} + \mathbf{c} \times \mathbf{a} + (\mathbf{c} - \mathbf{a}) \times (\mathbf{b} - \mathbf{a})] \\ &\quad = \tfrac{1}{2}[\mathbf{a} \times \mathbf{b} + \mathbf{b} \times \mathbf{c} + \mathbf{c} \times \mathbf{a} + \mathbf{c} \times \mathbf{b} - \mathbf{a} \times \mathbf{b} - \mathbf{c} \times \mathbf{a}] \\ &\quad = 0 \end{aligned}$$

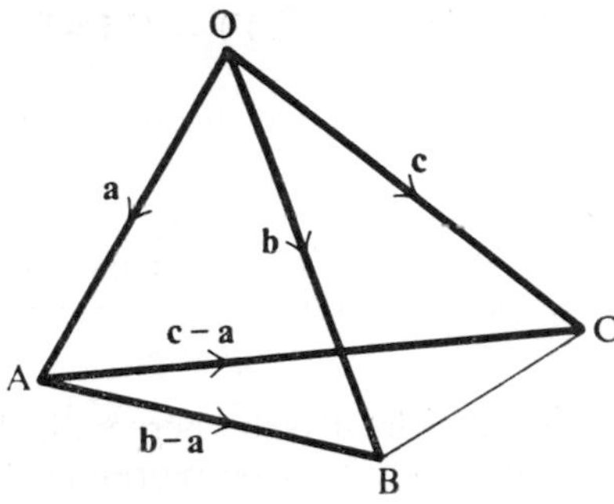

Figure 2.21

(E) Find the moment of a force $\mathbf{F} = X\mathbf{i} + Y\mathbf{j} + Z\mathbf{k}$, which passes through a point R (x, y, z), about the origin O (Fig. 2.22).

The position vector of R is $\mathbf{r} = x\mathbf{i} + y\mathbf{j} + z\mathbf{k}$.

The moment vector $\mathbf{G}$ of $\mathbf{F}$ about O is given by

$$\mathbf{G} = \mathbf{r} \times \mathbf{F} = \begin{vmatrix} \mathbf{i} & \mathbf{j} & \mathbf{k} \\ x & y & z \\ X & Y & Z \end{vmatrix}$$

$$= (yZ - zY)\mathbf{i} + (zX - xZ)\mathbf{j} + (xY - yX)\mathbf{k}$$

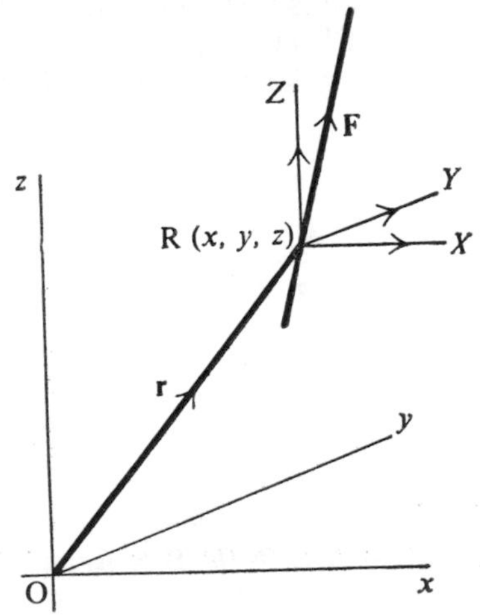

Figure 2.22

In particular, the moment of **F** about Ox is $(yZ - zY) = \mathbf{G.i}$ which is a scalar quantity.

Similarly $zX - xZ = \mathbf{G.j}$ and $(xY - yX) = \mathbf{G.k}$ are the moments of **F** about Oy and Oz respectively.

(F) A rigid body rotates with angular velocity ω about an axis. Find the vector velocity **v** of the point R of the body whose position vector with respect to a point O on the axis is **r**.

Solution Let ON be the axis of rotation and let RN be perpendicular to ON (Fig. 2.23). The angular velocity of the body may be represented by a

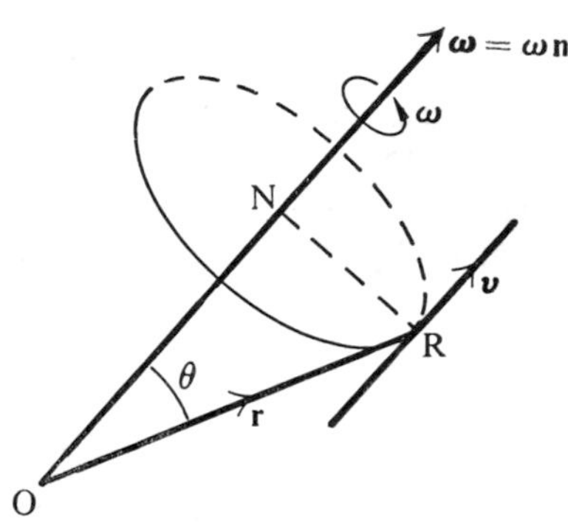

Figure 2.23

vector $\omega\mathbf{n}$ where **n** is a unit vector whose direction is related to the direction of rotation in accordance with the convention of Section 2.5.

R is moving in a circle of radius RN with angular velocity ω. Thus the magnitude of the velocity **v** of R is

$$\omega\text{RN} = \omega r \sin \theta$$

The direction of **v** is perpendicular to both **n** and **r** and is directed into the paper. Therefore

$$\mathbf{v} = \omega(\mathbf{n} \times \mathbf{r}) = \boldsymbol{\omega} \times \mathbf{r}$$

EXERCISES 2.3

1. If $\mathbf{A} = \mathbf{i} - 2\mathbf{j} + \mathbf{k}$ and $\mathbf{B} = 2\mathbf{i} + \mathbf{j} - 3\mathbf{k}$, find (*a*) $\mathbf{A} \times \mathbf{B}$, (*b*) $\mathbf{B} \times \mathbf{A}$, and show that $(\mathbf{B} - \mathbf{A}) \times (\mathbf{B} + \mathbf{A}) = -2(\mathbf{A} \times \mathbf{B})$.
   [(*a*) $5(\mathbf{i} + \mathbf{j} + \mathbf{k})$, (*b*) $-5(\mathbf{i} + \mathbf{j} + \mathbf{k})$]

2. If $\mathbf{A} = -\mathbf{i} + 2\mathbf{j} - \mathbf{k}$ and $\mathbf{B} = 2\mathbf{i} - \mathbf{j} + 3\mathbf{k}$, find (*a*) $\mathbf{A} \times \mathbf{B}$, (*b*) $|\mathbf{A} + \mathbf{B}|$, (*c*) the unit vector parallel to $\mathbf{A} \times \mathbf{B}$, (*d*) $(\mathbf{A} + 2\mathbf{B}) \times (\mathbf{A} - \mathbf{B})$.

   [(*a*) $5\mathbf{i} + \mathbf{j} - 3\mathbf{k}$, (*b*) $\sqrt{35}$, (*c*) $\dfrac{1}{\sqrt{35}}(5\mathbf{i} + \mathbf{j} - 3\mathbf{k})$, (*d*) $-3(5\mathbf{i} + \mathbf{j} - 3\mathbf{k})$]

3. The vertices of a triangle are at the points (5, −1, 1), (4, 1, −2) and (3, 0, 2). Find its area. [$\frac{1}{2}\sqrt{83}$]

4. Show that the perpendicular distance from the point A to the straight line joining points B and C is

$$\frac{|\mathbf{a} \times \mathbf{b} + \mathbf{b} \times \mathbf{c} + \mathbf{c} \times \mathbf{a}|}{|\mathbf{b} - \mathbf{c}|}$$

   where **a**, **b**, **c** are the position vectors of the points A, B, C respectively. Hence calculate the perpendicular distance of the point (−5, 2, 3) from the line joining the points (−1, 3, −4) and (2, 3, 4). [$\sqrt{(2\,882/73)}$]

5. A rigid body is rotating about an axis joining the origin to the point (6, −3, 2) with angular velocity 14 rad/s. Find the velocity vector of the point (4, 1, 3) of the body if distances are measured in metres. [$\pm 2(11\mathbf{i} + 10\mathbf{j} - 18\mathbf{k})$ m/s]

6. Show that three points A, B, C, having position vectors **a**, **b**, **c** respectively, will be collinear if

$$\mathbf{a} \times \mathbf{b} + \mathbf{b} \times \mathbf{c} + \mathbf{c} \times \mathbf{a} = 0$$

7. If a particle of mass m and charge e moves with velocity $\mathbf{q}$ in an electric field **E** and a magnetic field **H**, it experiences a force

$$\mathbf{F} = e\mathbf{E} + e\mathbf{q} \times \mathbf{H}$$

   If $\mathbf{q} = u\mathbf{i} + v\mathbf{j} + w\mathbf{k}$ and $\mathbf{E} = E\mathbf{j}$, $\mathbf{H} = H\mathbf{k}$, show that the equations of motion of the particle are

$$m\dot{u} = evH \qquad m\dot{v} = eE - euH \qquad m\dot{w} = 0$$

   where $\dot{u}$ and $\dot{v}$ are derivatives with respect to time.

Products of Three Vectors

Since $\mathbf{B} \times \mathbf{C}$ is a vector, the products $\mathbf{A}.(\mathbf{B} \times \mathbf{C})$ and $\mathbf{A} \times (\mathbf{B} \times \mathbf{C})$ each have a meaning; the first is a scalar and the second a vector.

2.8 Scalar Triple Product

The brackets in the scalar triple product $\mathbf{A}.(\mathbf{B} \times \mathbf{C})$ are often omitted since $(\mathbf{A}.\mathbf{B}) \times \mathbf{C}$ has no meaning and therefore $\mathbf{A}.\mathbf{B} \times \mathbf{C}$ is unambiguous.

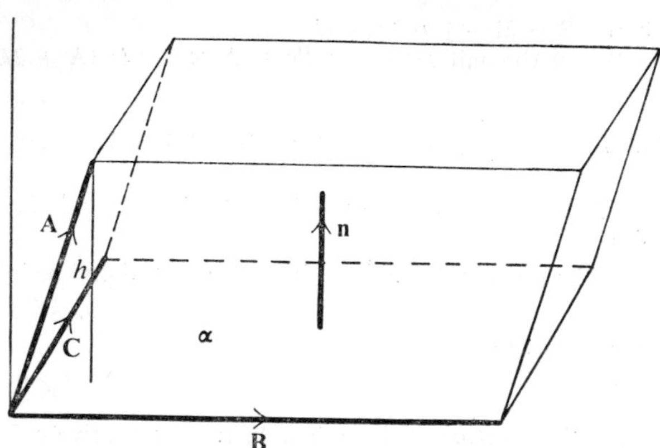

Figure 2.24

Figure 2.24 illustrates a parallelepiped whose sides represent the vectors **A**, **B**, **C** (shown as a right-handed system of vectors).

Let **n** be a unit vector normal to the parallelogram α formed by the vectors **B** and **C** and having the direction of **B** × **C**. Let h be the height of the parallelepiped in the direction of **n**. Volume of the parallelepiped is

$$V = \text{Height } h \times \text{Area of parallelogram } \alpha$$

$$= (\mathbf{A}.\mathbf{n})(|\mathbf{B} \times \mathbf{C}|) = \mathbf{A}.|\mathbf{B} \times \mathbf{C}|\,\mathbf{n} = \mathbf{A}.\mathbf{B} \times \mathbf{C}$$

If **A**, **B**, **C** do not form a right-handed system, **A.n** will be negative and hence **A.B** × **C** $= -V$. Therefore

$$\mathbf{A}.\mathbf{B} \times \mathbf{C} = \pm V$$

Similarly, **B.C** × **A** and **C.A** × **B** have the value $\pm V$, so that

$$\mathbf{A}.\mathbf{B} \times \mathbf{C} = \mathbf{B}.\mathbf{C} \times \mathbf{A} = \mathbf{C}.\mathbf{A} \times \mathbf{B} = -V \tag{2.22}$$

the + or − sign being taken according as **A**, **B**, **C** do or do not form a right-handed system.

From eqn (2.22), it follows that

$$\mathbf{A}.\mathbf{B} \times \mathbf{C} = \mathbf{C}.\mathbf{A} \times \mathbf{B} = \mathbf{A} \times \mathbf{B}.\mathbf{C} \tag{2.23}$$

so that in a scalar triple product, the dot and cross may be interchanged without changing its value. It now follows that

$$\mathbf{A.A \times C = A \times A.C} = 0$$

The notation [**A B C**] or [**A, B, C**] is often used to denote **A.B × C** or **A × B.C**.

If three vectors **A**, **B**, **C** are coplanar, the volume of the parallelepiped formed by them is zero. Thus **A.B × C** = 0. The converse is also true. Thus a necessary and sufficient condition that three vectors **A**, **B**, **C** be coplanar is that [**A B C**] = 0.

By eqn (2.18) we have

$$\mathbf{B \times C} = (B_2C_3 - B_3C_2)\mathbf{i} + (B_3C_1 - B_1C_3)\mathbf{j} + (B_1C_2 - B_2C_1)\mathbf{k}$$

Therefore

$$\mathbf{A.B \times C} = A_1(B_2C_3 - B_3C_2) + A_2(B_3C_1 - B_1C_3) + A_3(B_1C_2 - B_2C_1)$$

$$= \begin{vmatrix} A_1 & A_2 & A_3 \\ B_1 & B_2 & B_3 \\ C_1 & C_2 & C_3 \end{vmatrix} \tag{2.24}$$

Since the sign of a determinant is changed by interchanging two of its rows it follows that

$$\mathbf{A.B \times C} = - \begin{vmatrix} B_1 & B_2 & B_3 \\ A_1 & A_2 & A_3 \\ C_1 & C_2 & C_3 \end{vmatrix} = + \begin{vmatrix} B_1 & B_2 & B_3 \\ C_1 & C_2 & C_3 \\ A_1 & A_2 & A_3 \end{vmatrix} = \mathbf{B.C \times A}$$

as in eqn (2.22).

The *distributive law* holds for scalar triple products since it holds for both scalar and vector products, thus for example

$$[\mathbf{r, a - b, c - d}] = [\mathbf{r\ a\ c}] + [\mathbf{r\ b\ d}] - [\mathbf{r\ b\ c}] - [\mathbf{r\ a\ d}]$$

It is, of course, essential to preserve the order of the factors.

2.9 *Equation of the Plane through Three Non-collinear Points*

Let A, B, C be the three non-collinear points which define the plane and let R(x, y, z) be any point on it (Fig. 2.25). The vectors $(\mathbf{r} - \mathbf{a})$, $(\mathbf{b} - \mathbf{a})$,

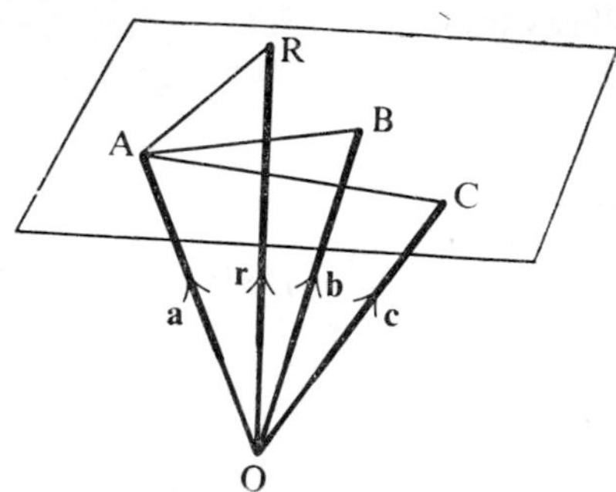

Figure 2.25

(**c** — **a**) are coplanar. Therefore

$$(\mathbf{r} - \mathbf{a}).(\mathbf{b} - \mathbf{a}) \times (\mathbf{c} - \mathbf{a}) = 0 \tag{2.25}$$

i.e. $(\mathbf{r} - \mathbf{a}).[\mathbf{b} \times \mathbf{c} + \mathbf{c} \times \mathbf{a} + \mathbf{a} \times \mathbf{b} + \mathbf{a} \times \mathbf{a}] = 0.$

$$\mathbf{r}.[\mathbf{a} \times \mathbf{b} + \mathbf{b} \times \mathbf{c} + \mathbf{c} \times \mathbf{a}] = \mathbf{a}.\mathbf{b} \times \mathbf{c} \tag{2.26}$$

If A, B, C are the points (x_r, y_r, z_r), $(r = 1, 2, 3)$ respectively, the equation (2.25) may be written in the determinantal form

$$\begin{vmatrix} (x - x_1) & (y - y_1) & (z - z_1) \\ (x_2 - x_1) & (y_2 - y_1) & (z_2 - z_1) \\ (x_3 - x_1) & (y_3 - y_1) & (z_3 - z_1) \end{vmatrix} = 0 \tag{2.27}$$

2.10 *Equation of the Plane through a Given Line and Parallel to Another Line*

Let the plane pass through the line $\mathbf{r} = \mathbf{a} + t\mathbf{b}$ and be parallel to the vector **c**. This plane contains the point A and is parallel to the two vectors **b** and **c**. It follows that the vector $\mathbf{b} \times \mathbf{c}$ is perpendicular to the plane. Therefore by eqn (2.2),

$$(\mathbf{r} - \mathbf{a}).\mathbf{b} \times \mathbf{c} = 0 \qquad \text{i.e. } [\mathbf{r}\ \mathbf{b}\ \mathbf{c}] = [\mathbf{a}\ \mathbf{b}\ \mathbf{c}] \tag{2.28}$$

2.11 *The Common Perpendicular to Two Skew Lines*

Let the equations of the two lines be

$$\mathbf{r} = \mathbf{a} + t\mathbf{b} \qquad \text{and} \qquad \mathbf{r} = \mathbf{c} + s\mathbf{d}$$

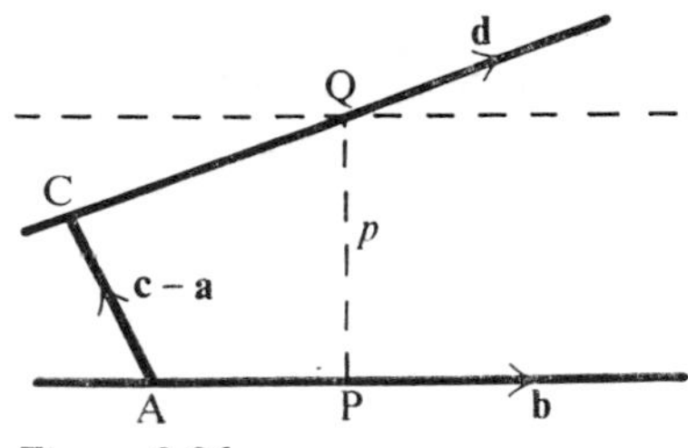

Figure 2.26

Let A and C be the points on these lines whose position vectors are **a** and **c**. Let PQ, of length p, be their common perpendicular.

Since PQ is perpendicular to both **b** and **d**, then PQ is parallel to $\mathbf{N} = \mathbf{b} \times \mathbf{d}$.

The magnitude of PQ is equal to the magnitude of the projection **AC** on **N**. Therefore

$$p = \frac{1}{N}(\mathbf{c} - \mathbf{a}).\mathbf{N} = \frac{1}{N}(\mathbf{c} - \mathbf{a}).\mathbf{b} \times \mathbf{d} \tag{2.29}$$

It follows that the two lines intersect if $(\mathbf{c} - \mathbf{a}).\mathbf{b} \times \mathbf{d} = 0$ which is otherwise obvious since this is the condition that $(\mathbf{c} - \mathbf{a})$, **b** and **d** should be coplanar vectors.

The equation of the plane through the line AP and the common perpendicular PQ is

$$(\mathbf{r} - \mathbf{a}).\mathbf{b} \times \mathbf{N} = 0 \qquad \text{i.e. } (\mathbf{r} - \mathbf{a}).\mathbf{b} \times (\mathbf{b} \times \mathbf{d}) = 0$$

The plane through CQ and PQ is

$$(\mathbf{r} - \mathbf{c}).\mathbf{d} \times (\mathbf{b} \times \mathbf{d}) = 0$$

These two planes determine the common perpendicular which is their line of intersection.

EXAMPLE 2.8

Find the volume of the parallepiped whose three concurrent sides are the vectors

$$\mathbf{A} = 3\mathbf{i} - \mathbf{j} + \mathbf{k} \qquad \mathbf{B} = \mathbf{i} + 2\mathbf{j} - 3\mathbf{k} \qquad \mathbf{C} = -2\mathbf{i} + 5\mathbf{j}$$

$$\pm V = \mathbf{C}.\mathbf{A} \times \mathbf{B} = \begin{vmatrix} -2 & 5 & 0 \\ 3 & -1 & 1 \\ 1 & 2 & -3 \end{vmatrix} = -2(1) - 5(-10) = 48$$

EXAMPLE 2.9

Two straight lines pass through the points A(5, 1, 2) and B(3, 0, 1) having the directions of the vectors $\mathbf{c} = 2\mathbf{i} - \mathbf{j} + 3\mathbf{k}$ and $\mathbf{d} = -\mathbf{i} + 2\mathbf{j} - 2\mathbf{k}$ respectively. Find the shortest distance between these lines, and the coordinates of the feet of the common perpendicular.

Solution $\mathbf{c} \times \mathbf{d}$ is parallel to the common perpendicular to the two lines, and

$$\mathbf{c} \times \mathbf{d} = \begin{vmatrix} \mathbf{i} & \mathbf{j} & \mathbf{k} \\ 2 & -1 & 3 \\ -1 & 2 & -2 \end{vmatrix} = -4\mathbf{i} + \mathbf{j} + 3\mathbf{k}$$

A unit vector $\mathbf{n}$ in the direction of the common perpendicular is

$$\mathbf{n} = \frac{-4\mathbf{i} + \mathbf{j} + 3\mathbf{k}}{\sqrt{26}}$$

Shortest distance between the lines is

$$p = (\mathbf{a} - \mathbf{b}).\mathbf{n} = \frac{(2\mathbf{i} + \mathbf{j} + \mathbf{k}).(-4\mathbf{i} + \mathbf{j} + 3\mathbf{k})}{\sqrt{26}} = 4/\sqrt{26}$$

Let the common perpendicular PQ meet the lines through A and B in P and Q respectively.

The equation of the plane through AP and PQ is $[\mathbf{r} - \mathbf{a}, \mathbf{c}, \mathbf{n}] = 0$, i.e.

$$\begin{vmatrix} (x-5) & (y-1) & (z-2) \\ 2 & -1 & 3 \\ 4 & -1 & -3 \end{vmatrix} = 0$$

$$6(x-5) + 18(y-1) + 2(z-2) = 0 \qquad \text{i.e. } 3x + 9y + z = 26$$

The equation of the straight line BQ is

$$\frac{x-3}{-1} = \frac{y}{2} = \frac{z-1}{-2} = k$$

This straight line meets the plane in Q for which

$$9 - 3k + 18k + 1 - 2k = 26 \qquad \text{i.e. } k = 16/13$$

Therefore Q is the point $\frac{1}{13}(23, 32, -19)$.

Similarly, P may be shown to be the point $\frac{1}{13}(31, 30, -25)$.

EXAMPLE 2.10
Find the volume of a tetrahedron.

Solution Let p be the magnitude of the common perpendicular to the two

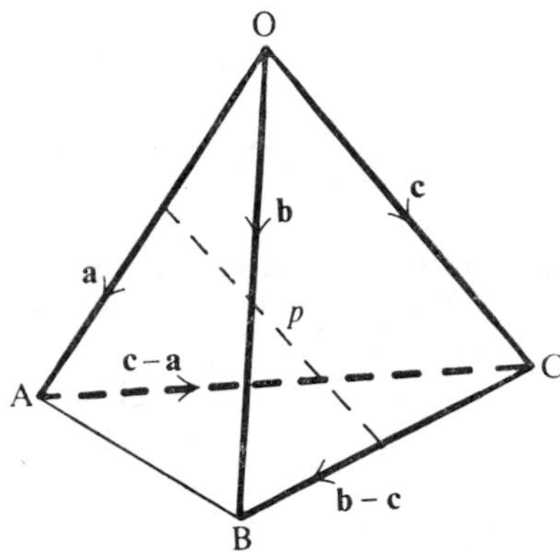

Figure 2.27

opposite edges OA, BC of a tetrahedron (Fig. 2.27). Let **n** be a unit vector parallel to the common perpendicular. Then **n** will be perpendicular to both **a** and (**b** − **c**). Thus

$$\mathbf{n} = \frac{\mathbf{a} \times (\mathbf{b} - \mathbf{c})}{\mathrm{OA.BC.}\sin \alpha}$$

where α is the angle of inclination of OA to BC.

Now p is the projection of CA on **n**. Therefore

$$p = (\mathbf{c} - \mathbf{a}).\mathbf{n} = \frac{(\mathbf{c} - \mathbf{a}).\mathbf{a} \times (\mathbf{b} - \mathbf{c})}{\mathrm{OA.BC.}\sin \alpha}$$

$$= \frac{[\mathbf{a}, \mathbf{b}, \mathbf{c}]}{\mathrm{OA.BC.}\sin \alpha} \quad \text{on expanding the numerator.}$$

The volume V of the tetrahedron is given by

$$V = \tfrac{1}{3}\mathbf{a}.\tfrac{1}{2}(\mathbf{b} \times \mathbf{c}) = \tfrac{1}{6}\,[\mathbf{a}, \mathbf{b}, \mathbf{c}] = {}^{1}\mathrm{OA.BC.}p \sin \alpha$$

EXERCIS S 2.4

1. Find the equation of the plane through the points (2, 0, 1), (3, 1, 5) and (−1, 2, −1). [$2x + 2y - z = 3$]

2. Show that the equation of the plane through the point A parallel to each of the vectors $\mathbf{b}$ and $\mathbf{c}$ is $[\mathbf{r}\,\mathbf{b}\,\mathbf{c}] = [\mathbf{a}\,\mathbf{b}\,\mathbf{c}]$.
   Hence, or otherwise, find the cartesian form of the equation of the plane through the point $(-2, 1, 3)$ parallel to the vectors $-\mathbf{i} + 2\mathbf{k}$ and $3\mathbf{i} + 2\mathbf{j} - \mathbf{k}$. $[4x - 5y + 2z + 7 = 0]$
3. Show that the equation of the plane through the points A and B parallel to a vector $\mathbf{c}$ is $[\mathbf{r}, (\mathbf{b} - \mathbf{a}), \mathbf{c}] = [\mathbf{a}\,\mathbf{b}\,\mathbf{c}]$.
   Hence, or otherwise, find the cartesian form of the equation of the plane through the points $(5, 1, 2)$ and $(2, 1, 0)$ parallel to the vector $3\mathbf{i} + \mathbf{j} - 4\mathbf{k}$. $[2x - 18y - 3z + 14 = 0]$
4. Show that the line through A $(-5, -8, -3)$ and B $(2, 13, 11)$ intersects the line through C $(-7, 0, -5)$ and D $(5, -6, 13)$. What is the point of intersection? What is the point of intersection of AC and BD? $[(-3, -2, 1); (23, -120, 25)]$
5. Show that the shortest distance between the straight line through the points $(4, 2, 5)$ and $(-3, -1, 2)$ and the straight line through the points $(6, -3, -1)$ and $(-4, 3, -5)$ is $241/\sqrt{1\,522}$.
6. Show that the equation of the plane which is perpendicular to the plane $\mathbf{r}.\mathbf{a}$ = constant and passes through the line $\mathbf{r} = \mathbf{b} + t\mathbf{c}$ is $[\mathbf{r}\,\mathbf{c}\,\mathbf{a}] = [\mathbf{b}\,\mathbf{c}\,\mathbf{a}]$.
7. Show that the equation of the straight line drawn through the point whose position vector is $\mathbf{p}$ to intersect both of the lines $\mathbf{r} = \mathbf{a} + t\mathbf{b}$ and $\mathbf{r} = \mathbf{c} + s\mathbf{d}$ is $\mathbf{r} = \mathbf{p} + \mathbf{k}.(\mathbf{n}_1 \times \mathbf{n}_2)$ where $\mathbf{n}_1 = \mathbf{b} \times (\mathbf{a} - \mathbf{p})$ and $\mathbf{n}_2 = \mathbf{d} \times (\mathbf{c} - \mathbf{p})$.
8. Show that

$$[\mathbf{A}\,\mathbf{B}\,\mathbf{C}][\mathbf{a}\,\mathbf{b}\,\mathbf{c}] = \begin{vmatrix} \mathbf{A}.\mathbf{a} & \mathbf{A}.\mathbf{b} & \mathbf{A}.\mathbf{c} \\ \mathbf{B}.\mathbf{a} & \mathbf{B}.\mathbf{b} & \mathbf{B}.\mathbf{c} \\ \mathbf{C}.\mathbf{a} & \mathbf{C}.\mathbf{b} & \mathbf{C}.\mathbf{c} \end{vmatrix}$$

9. Two forces $\mathbf{F}_1 = \mathbf{i} + \mathbf{j} + \mathbf{k}$ and $\mathbf{F}_2 = \mathbf{i} + 2\mathbf{j} - \mathbf{k}$ act through points whose position vectors are

$$\mathbf{S}_1 = \mathbf{i} + \mathbf{j} + 2\mathbf{k} \quad \text{and} \quad \mathbf{S}_2 = q\mathbf{j} + 5\mathbf{k}$$

respectively, relative to a fixed point and in terms of three mutually perpendicular unit vectors $\mathbf{i}$, $\mathbf{j}$ and $\mathbf{k}$. If the lines of action of $\mathbf{F}_1$ and $\mathbf{F}_2$ intersect, find q and find the vector equation of the line of action of the resultant of $\mathbf{F}_1$ and $\mathbf{F}_2$. (*U.L.*) $[q = -2; \mathbf{r} = 2(1 + t)\mathbf{i} + (2 + 3t)\mathbf{j} + 3\mathbf{k}]$

2.12 Vector Triple Product

The vector triple product

$$\mathbf{T} = \mathbf{A} \times (\mathbf{B} \times \mathbf{C})$$

is a vector which is perpendicular to each of the vectors $\mathbf{A}$ and $(\mathbf{B} \times \mathbf{C})$. Since $(\mathbf{B} \times \mathbf{C})$ is a vector normal to the plane of $\mathbf{B}$ and $\mathbf{C}$, it follows that $\mathbf{T}$

ies in the plane of **B** and **C** and must therefore be expressible in the form

$$\mathbf{T} = \alpha\mathbf{B} + \beta\mathbf{C}$$

Choose coordinate axes, so that **B** and **C** lie in the plane of Ox, Oy and Ox has the direction of **B** (Fig. 2.28). Therefore

$$\mathbf{B} = B\mathbf{i} \qquad \mathbf{C} = C_1\mathbf{i} + C_2\mathbf{j}$$

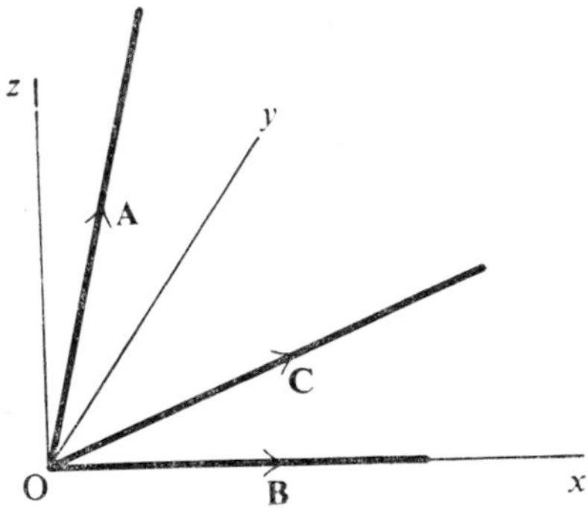

Figure 2.28

Let $\mathbf{A} = \mathrm{A}_1\mathbf{i} + A_2\mathbf{j} + A_3\mathbf{k}$. Thus, $\mathbf{B} \times \mathbf{C} = BC_2\mathbf{k}$ and

$$\mathbf{A} \times (\mathbf{B} \times \mathbf{C}) = \begin{vmatrix} \mathbf{i} & \mathbf{j} & \mathbf{k} \\ A_1 & A_2 & A_3 \\ 0 & 0 & BC_2 \end{vmatrix}$$

$$= (A_2C_2B)\mathbf{i} - (A_1C_2B)\mathbf{j}$$

$$= (A_1C_1 + A_2C_2)B\mathbf{i} - A_1B(C_1\mathbf{i} + C_2\mathbf{j})$$

i.e.

$$\mathbf{A} \times (\mathbf{B} \times \mathbf{C}) = (\mathbf{A}.\mathbf{C})\mathbf{B} - (\mathbf{A}.\mathbf{B})\mathbf{C} \tag{2.30}$$

It follows that

$$(\mathbf{A} \times \mathbf{B}) \times \mathbf{C} = -\mathbf{C} \times (\mathbf{A} \times \mathbf{B}) = -(\mathbf{C}.\mathbf{B})\mathbf{A} + (\mathbf{C}.\mathbf{A})\mathbf{B} \tag{2.31}$$

so that $(\mathbf{A} \times \mathbf{B}) \times \mathbf{C} \neq \mathbf{A} \times (\mathbf{B} \times \mathbf{C})$.

Thus, if the position of the bracket in a vector triple product is changed, the value of the product is altered, i.e. the associative law is not, in general valid for vector products.

Products of Four Vectors

The scalar product $(\mathbf{A} \times \mathbf{B}).(\mathbf{C} \times \mathbf{D})$ and the vector product $(\mathbf{A} \times \mathbf{B}) \times (\mathbf{C} \times \mathbf{D})$ of four vectors occur, not infrequently, in vector analysis. Each is readily expressible in terms of scalar products.

2.13 Scalar Product of Four Vectors

By eqn (2.23), we have

$$\mathbf{P}.\mathbf{C} \times \mathbf{D} = \mathbf{P} \times \mathbf{C}.\mathbf{D}$$

Now let $\mathbf{P} = \mathbf{A} \times \mathbf{B}$, then

$$\begin{aligned}(\mathbf{A} \times \mathbf{B}).(\mathbf{C} \times \mathbf{D}) &= \{(\mathbf{A} \times \mathbf{B}) \times \mathbf{C}\}.\mathbf{D} \\ &= \{(\mathbf{A}.\mathbf{C})\mathbf{B} - (\mathbf{B}.\mathbf{C})\mathbf{A}\}.\mathbf{D} \quad \text{by eqn. (2.31)} \\ &= (\mathbf{A}.\mathbf{C})(\mathbf{B}.\mathbf{D}) - (\mathbf{B}.\mathbf{C})(\mathbf{A}.\mathbf{D}) \qquad (2.32)\end{aligned}$$

2.14 Vector Product of Four Vectors

Assume that the four vectors are localized at a point. The vector $(\mathbf{A} \times \mathbf{B}) \times (\mathbf{C} \times \mathbf{D})$ is perpendicular to the vector $(\mathbf{A} \times \mathbf{B})$ and therefore lies in the plane of $\mathbf{A}$ and $\mathbf{B}$. It is therefore expressible as a linear function of $\mathbf{A}$ and $\mathbf{B}$.

Similarly, this vector lies in the plane of $\mathbf{C}$ and $\mathbf{D}$ and is expressible as a linear function of $\mathbf{C}$ and $\mathbf{D}$. It must clearly have the direction of the line of intersection of these two planes.

Now $\mathbf{P} \times (\mathbf{C} \times \mathbf{D}) = (\mathbf{P}.\mathbf{D})\mathbf{C} - (\mathbf{P}.\mathbf{C})\mathbf{D}$ by eqn (2.30)

Let $\mathbf{P} = (\mathbf{A} \times \mathbf{B})$, then

$$\begin{aligned}(\mathbf{A} \times \mathbf{B}) \times (\mathbf{C} \times \mathbf{D}) &= \{\mathbf{A} \times \mathbf{B}.\mathbf{D}\}\mathbf{C} - \{\mathbf{A} \times \mathbf{B}.\mathbf{C}\}\mathbf{D} \\ &= [\mathbf{A}\ \mathbf{B}\ \mathbf{D}]\mathbf{C} - [\mathbf{A}\ \mathbf{B}\ \mathbf{C}]\mathbf{D} \qquad (2.33)\end{aligned}$$

Similarly

$$(\mathbf{A} \times \mathbf{B}) \times \mathbf{Q} = (\mathbf{A}.\mathbf{Q})\mathbf{B} - (\mathbf{B}.\mathbf{Q})\mathbf{A}$$

$$(\mathbf{A} \times \mathbf{B}) \times (\mathbf{C} \times \mathbf{D}) = [\mathbf{A}\ \mathbf{C}\ \mathbf{D}]\mathbf{B} - [\mathbf{B}\ \mathbf{C}\ \mathbf{D}]\mathbf{A} \tag{2.34}$$

Equating the two expressions in (2.33) and (2.34) we have the following relation between any four vectors **A**, **B**, **C**, **D**:

$$[\mathbf{B}\ \mathbf{C}\ \mathbf{D}]\mathbf{A} - [\mathbf{A}\ \mathbf{C}\ \mathbf{D}]\mathbf{B} + [\mathbf{A}\ \mathbf{B}\ \mathbf{D}]\mathbf{C} - [\mathbf{A}\ \mathbf{B}\ \mathbf{C}]\mathbf{D} = 0 \tag{2.35}$$

[cf. eqn (1.7), linear dependence of four vectors]
Thus

$$\mathbf{D} = \frac{[\mathbf{B}\ \mathbf{C}\ \mathbf{D}]\mathbf{A} - [\mathbf{A}\ \mathbf{C}\ \mathbf{D}]\mathbf{B} + [\mathbf{A}\ \mathbf{B}\ \mathbf{D}]\mathbf{C}}{[\mathbf{A}\ \mathbf{B}\ \mathbf{C}]} \tag{2.36}$$

showing that any vector **D** may be expressed as a linear function of three other vectors **A**, **B**, **C** provided $[\mathbf{A}\,\mathbf{B}\,\mathbf{C}] \neq 0$, i.e. provided **A**, **B**, **C** are not coplanar.

EXAMPLE 2.11

(A) By expressing the vectors in terms of their rectangular components, show that

$$\mathbf{A} \times (\mathbf{B} \times \mathbf{C}) = (\mathbf{A}.\mathbf{C})\mathbf{B} - (\mathbf{A}.\mathbf{B})\mathbf{C}$$

Solution

$$\begin{aligned}
\mathbf{A} \times (\mathbf{B} \times \mathbf{C}) &= (A_1\mathbf{i} + A_2\mathbf{j} + A_3\mathbf{k}) \times [(B_2C_3 - B_3C_2)\mathbf{i} \\
&\quad + (B_3C_1 - B_1C_3)\mathbf{j} + (B_1C_2 - B_2C_1)\mathbf{k}] \\
&= [A_2(B_1C_2 - B_2C_1) - A_3(B_3C_1 - B_1C_3)]\mathbf{i} \\
&\quad + [A_3(B_2C_3 - B_3C_2) - A_1(B_1C_2 - B_2C_1)]\mathbf{j} \\
&\quad + [A_1(B_3C_1 - B_1C_3) - A_2(B_2C_3 - B_3C_2)]\mathbf{k} \\
&= (A_1C_1 + A_2C_2 + A_3C_3)(B_1\mathbf{i} + B_2\mathbf{j} + B_3\mathbf{k}) \\
&\quad - (A_1B_1 + A_2B_2 + A_3B_3)(C_1\mathbf{i} + C_2\mathbf{j} + C_3\mathbf{k}) \\
&= (\mathbf{A}.\mathbf{C})\mathbf{B} - (\mathbf{A}.\mathbf{B})\mathbf{C}
\end{aligned}$$

(B) Prove that $(\mathbf{A} \times \mathbf{B}).(\mathbf{B} \times \mathbf{C}) \times (\mathbf{C} \times \mathbf{A}) = (\mathbf{A}.\mathbf{B} \times \mathbf{C})^2$

Solution By eqn (2.33)

$$(\mathbf{B} \times \mathbf{C}) \times (\mathbf{C} \times \mathbf{A}) = [\mathbf{B}\ \mathbf{C}\ \mathbf{A}]\mathbf{C} - [\mathbf{B}\ \mathbf{C}\ \mathbf{C}]\mathbf{A}$$
$$= [\mathbf{A}\ \mathbf{B}\ \mathbf{C}]\mathbf{C}$$
$$(\mathbf{A} \times \mathbf{B}).(\mathbf{B} \times \mathbf{C}) \times (\mathbf{C} \times \mathbf{A}) = [\mathbf{A}\ \mathbf{B}\ \mathbf{C}](\mathbf{A} \times \mathbf{B}.\mathbf{C}) = [\mathbf{A}\ \mathbf{B}\ \mathbf{C}]^2$$

EXAMPLE 2.12 RECIPROCAL GROUPS OF VECTORS
From eqn (2.36), it is clear that any vector **r** may be expressed in terms of three non-coplanar vectors **a, b, c** ($[\mathbf{a}\ \mathbf{b}\ \mathbf{c}] \neq 0$) as follows:

$$\mathbf{r} = \frac{[\mathbf{r}\ \mathbf{b}\ \mathbf{c}]\mathbf{a} + [\mathbf{r}\ \mathbf{c}\ \mathbf{a}]\mathbf{b} + [\mathbf{r}\ \mathbf{a}\ \mathbf{b}]\mathbf{c}}{[\mathbf{a}\ \mathbf{b}\ \mathbf{c}]}$$
$$= (\mathbf{r}.\mathbf{a}')\mathbf{a} + (\mathbf{r}.\mathbf{b}')\mathbf{b} + (\mathbf{r}.\mathbf{c}')\mathbf{c}$$

where $\mathbf{a}' = \mathbf{b} \times \mathbf{c}/[\mathbf{a}\ \mathbf{b}\ \mathbf{c}]$, $\mathbf{b}' = \mathbf{c} \times \mathbf{a}/[\mathbf{a}\ \mathbf{b}\ \mathbf{c}]$, $\mathbf{c}' = \mathbf{a} \times \mathbf{b}/[\mathbf{a}\ \mathbf{b}\ \mathbf{c}]$.
(**a, b, c**) and (**a′, b′, c′**) are said to be *reciprocal* groups of vectors.

The following relations exist between the two groups:

(i) $\mathbf{a}.\mathbf{a}' = \dfrac{\mathbf{a}.\mathbf{b} \times \mathbf{c}}{[\mathbf{a}\ \mathbf{b}\ \mathbf{c}]} = 1 = \mathbf{b}.\mathbf{b}' = \mathbf{c}.\mathbf{c}'$

(ii) the scalar product of any other pair, drawn one from each group, is zero, e.g. $\mathbf{b}.\mathbf{a}' = \dfrac{\mathbf{b}.\mathbf{b} \times \mathbf{c}}{[\mathbf{a}\ \mathbf{b}\ \mathbf{c}]} = 0.$

(iii) if $[\mathbf{a}\ \mathbf{b}\ \mathbf{c}] = V \neq 0$, then

$$[\mathbf{a}'\ \mathbf{b}'\ \mathbf{c}'] = \frac{(\mathbf{b} \times \mathbf{c}).(\mathbf{c} \times \mathbf{a}) \times (\mathbf{a} \times \mathbf{b})}{V^3} = \frac{[\mathbf{a}\ \mathbf{b}\ \mathbf{c}]^2}{V^3} \quad \text{(by Example 2.11B)}$$
$$= \frac{1}{V} \neq 0$$

Thus if **a, b, c** are non-coplanar, **a′, b′, c′** are also non-coplanar.

EXAMPLE 2.13 SPHERICAL TRIGONOMETRY
Let ABC be a spherical triangle on a sphere of unit radius (Fig. 2.29). By definition, the sides of this triangle are arcs of great circles of the sphere.

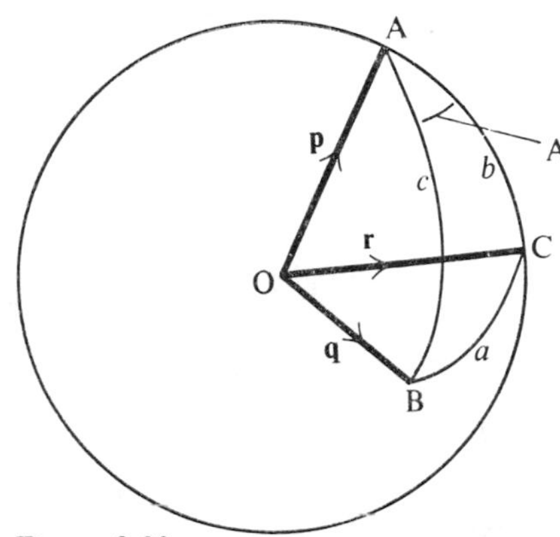

Figure 2.29

Let **p**, **q**, **r** be the position vectors of A, B, C with respect to O, the centre of the sphere. The side *a* of the triangle is the angle BÔC between the unit vectors **q** and **r** whilst the angle A of the triangle is the angle between the planes AOB and AOC. The other sides and angles are similarly interpreted.

By eqn (2.32)

$$(\mathbf{q} \times \mathbf{p}).(\mathbf{r} \times \mathbf{p}) = (\mathbf{q}.\mathbf{r})(\mathbf{p}.\mathbf{p}) - (\mathbf{q}.\mathbf{p})(\mathbf{p}.\mathbf{r})$$

Therefore

$$\sin c \sin b\, \mathbf{n}_1.\mathbf{n}_2 = \cos a.1 - \cos c.\cos b$$

where $\mathbf{n}_1$ is a unit vector normal to **q** and **p** drawn out of the paper and $\mathbf{n}_2$ is a unit vector normal to **r** and **p** drawn out of the paper. Thus the angle between $\mathbf{n}_1$ and $\mathbf{n}_2$ is A. Therefore

$$\mathbf{n}_1.\mathbf{n}_2 = \cos A \quad \text{and therefore}$$

$$\cos a = \cos b \cos c + \sin b \sin c \cos A$$

There are, of course, two similar formulae for $\cos b$ and $\cos c$.

Again we have by eqn (2.33),

$$(\mathbf{r} \times \mathbf{p}) \times (\mathbf{r} \times \mathbf{q}) = (\mathbf{r}.\mathbf{p} \times \mathbf{q})\mathbf{r}$$

Noting that **r** is a unit vector perpendicular to both $(\mathbf{r} \times \mathbf{p})$ and $(\mathbf{r} \times \mathbf{q})$, the above equation becomes

$$\sin b\mathbf{n}_2 \times \sin a\mathbf{n}_3 = [\mathbf{p}\ \mathbf{q}\ \mathbf{r}]\mathbf{r}$$

where $\mathbf{n}_3$ is a unit vector normal to $\mathbf{r}$ and $\mathbf{q}$

i.e. $\sin a \sin b \sin C\mathbf{r} = [\mathbf{p}\ \mathbf{q}\ \mathbf{r}]\mathbf{r}$ $\qquad \sin a \sin b \sin C = [\mathbf{p}\ \mathbf{q}\ \mathbf{r}]$

Cyclic permutations of $\mathbf{p}$, $\mathbf{q}$, $\mathbf{r}$ and the sides and angles give

$$\sin b \sin c \sin A = \sin c \sin a \sin B = [\mathbf{p}\ \mathbf{q}\ \mathbf{r}]$$

Hence $\sin A/\sin a = \sin B/\sin b = \sin C/\sin c$.

EXERCISES 2.5

1. Prove that $\mathbf{A} \times (\mathbf{B} \times \mathbf{C}) + \mathbf{B} \times (\mathbf{C} \times \mathbf{A}) + \mathbf{C} \times (\mathbf{A} \times \mathbf{B}) = 0$.
2. A vector $\mathbf{r}$ is resolved into two components, one parallel to $\mathbf{c}$ and the other perpendicular to $\mathbf{c}$. Show that the latter is

$$\frac{1}{c^2}(\mathbf{c} \times \mathbf{r}) \times \mathbf{c}$$

3. Show that
   (i) $(\mathbf{A} \times \mathbf{B}) \times \mathbf{A} = \mathbf{A} \times (\mathbf{B} \times \mathbf{A})$
   (ii) $\mathbf{A} \times (\mathbf{B} \times \mathbf{A}) \times (\mathbf{A} \times \mathbf{C}) = 0$
   (iii) $[\mathbf{A} \times \mathbf{B}, \mathbf{A} \times \mathbf{C}, \mathbf{D}] = [\mathbf{A}\ \mathbf{B}\ \mathbf{C}](\mathbf{A}.\mathbf{D})$
   (iv) $(\mathbf{A} \times \mathbf{B}).(\mathbf{C} \times \mathbf{D}) + (\mathbf{B} \times \mathbf{C}).(\mathbf{A} \times \mathbf{D}) + (\mathbf{C} \times \mathbf{A}).(\mathbf{B} \times \mathbf{D}) = 0$
4. Show that $(\mathbf{A} \times \mathbf{C}) \times \mathbf{B} = 0$ is a necessary and sufficient condition that $(\mathbf{A} \times \mathbf{B}) \times \mathbf{C} = \mathbf{A} \times (\mathbf{B} \times \mathbf{C})$. Discuss the cases where either $\mathbf{A}.\mathbf{B} = 0$ or $\mathbf{B}.\mathbf{C} = 0$.
5. Prove that

$$(\mathbf{a} \times \mathbf{b}) \times (\mathbf{c} \times \mathbf{d}) + (\mathbf{a} \times \mathbf{c}) \times (\mathbf{b} \times \mathbf{d}) = [\mathbf{a}\ \mathbf{c}\ \mathbf{d}]\mathbf{b} + [\mathbf{a}\ \mathbf{b}\ \mathbf{d}]\mathbf{c}$$

   Find the values of $\mathbf{r}$ and λ which satisfy the equations

$$\mathbf{r} \times \mathbf{a} = \mathbf{b} - \lambda\mathbf{a} \qquad \mathbf{r}.\mathbf{a} = 0$$

   where $\mathbf{a}$ and $\mathbf{b}$ are given vectors.
   Show that $[\mathbf{r}\ \mathbf{a}\ \mathbf{b}] = b^2 - (\mathbf{a}.\mathbf{b})^2/a^2$

$$\left[\mathbf{r} = \frac{1}{a^2}(\mathbf{a} \times \mathbf{b});\ \lambda = \frac{1}{a^2}(\mathbf{a}.\mathbf{b})\right]$$

6. Show that the two straight lines

$$\mathbf{r} = \mathbf{a} + k\mathbf{u} \qquad \mathbf{r} = \mathbf{b} + h\mathbf{v}$$

intersect if $[\mathbf{v\ a\ u}] = [\mathbf{v\ b\ u}]$ and that the point of intersection is expressible in the equivalent forms

$$\mathbf{a} + \frac{[\mathbf{a\ b\ v}]}{[\mathbf{v\ a\ u}]}\mathbf{u} \quad \text{or} \quad \mathbf{b} + \frac{[\mathbf{a\ b\ u}]}{[\mathbf{v\ b\ u}]}\mathbf{v}$$

7. With reference to Example 2.12, show that

$$\mathbf{a} = \frac{\mathbf{b}' \times \mathbf{c}'}{[\mathbf{a}'\ \mathbf{b}'\ \mathbf{c}']} \text{ and hence that } \mathbf{r} = (\mathbf{r.a})\mathbf{a}' + (\mathbf{r.b})\mathbf{b}' + (\mathbf{r.c})\mathbf{c}'$$

8. Find the set of vectors which are reciprocals of the set

$$3\mathbf{i} - \mathbf{j} + \mathbf{k} \qquad \mathbf{i} + \mathbf{j} - 2\mathbf{k} \qquad 2\mathbf{i} - 3\mathbf{j} - \mathbf{k}$$

$$\left[\frac{1}{23}(7\mathbf{i} + 3\mathbf{j} + 5\mathbf{k}), \frac{1}{23}(4\mathbf{i} + 5\mathbf{j} - 7\mathbf{k}), -\frac{1}{23}(\mathbf{i} + 7\mathbf{j} + 4\mathbf{k})\right]$$

3 Differentiation of vectors

3.1 Scalar and Vector Fields

If $\phi(x, y, z)$ has a scalar value at each point (x, y, z) of a region, then ϕ is said to be a *scalar function of position* and defines a *scalar field* ϕ in that region. For example, if the electrical potential in a region is given by the scalar function $\phi = xy^2z - x^2y$, then ϕ defines a scalar field.

On the other hand, if a vector $\mathbf{A}(x, y, z) = A_1(x, y, z)\mathbf{i} + A_2(x, y, z)\mathbf{j} + A_3(x, y, z)\mathbf{k}$ corresponds to each point of a region, $\mathbf{A}(x, y, z)$ is said to be a *vector function of position* and defines a *vector field* in that region. For example, if the velocity $\mathbf{v}$ in a fluid is given by $\mathbf{v} = xy\mathbf{i} + x^2\mathbf{j} - y^2z\mathbf{k}$, then $\mathbf{v}$ defines a vector field in the fluid.

3.2 Derivative of a Vector

Let $\mathbf{A}(u)$ be a vector function of a single scalar variable u. Let δu be an increment in u. The corresponding increment $\delta\mathbf{A}$ in $\mathbf{A}$ is given by (see Fig. 3.1)

$$\delta\mathbf{A} = \mathbf{A}(u + \delta u) - \mathbf{A}(u)$$

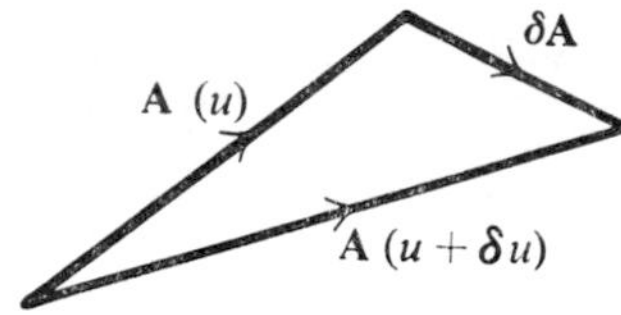

Figure 3.1

The derivative or differential coefficient of **A** with respect to u is defined to be

$$\lim_{\delta u \to 0} \frac{\delta \mathbf{A}}{\delta u}$$

(if this exists). It is denoted by $\dfrac{d\mathbf{A}}{du}$ and is itself a vector. Thus

$$\frac{d\mathbf{A}}{du} = \lim_{\delta u \to 0} \frac{\delta \mathbf{A}}{\delta u} = \lim_{\delta u \to 0} \frac{\mathbf{A}(u + \delta u) - \mathbf{A}(u)}{\delta u}$$

if this limit exists.

Since $d\mathbf{A}/du$ is a vector function of u,

$$\frac{d}{du}\left(\frac{d\mathbf{A}}{du}\right) \quad \text{or} \quad \frac{d^2A}{du^2}$$

may also be found if it exists. Similarly derivatives of higher orders may be found.

(Questions of continuity and differentiability will not be discussed here. It will be assumed that all the functions considered are continuous, single-valued and differentiable to any order.)

It is obvious that the derivative of any vector whose magnitude and direction are constant is zero. For example

$$\frac{d\mathbf{i}}{du} = \frac{d\mathbf{j}}{du} = \frac{d\mathbf{k}}{du} = 0$$

If $\mathbf{A}(u)$ and $\mathbf{B}(u)$ are vector functions of u, it may easily be shown that

$$\frac{d}{du}(\mathbf{A} + \mathbf{B}) = \frac{d\mathbf{A}}{du} + \frac{d\mathbf{B}}{du}$$

for if $\delta\mathbf{A}$ and $\delta\mathbf{B}$ are the increments of **A** and **B** corresponding to the increment δu,

$$\begin{aligned}\delta(\mathbf{A} + \mathbf{B}) &= (\mathbf{A} + \delta\mathbf{A} + \mathbf{B} + \delta\mathbf{B}) - (\mathbf{A} + \mathbf{B}) \\ &= \delta\mathbf{A} + \delta\mathbf{B}\end{aligned}$$

$$\frac{\delta(\mathbf{A} + \mathbf{B})}{\delta u} = \frac{\delta\mathbf{A}}{\delta u} + \frac{\delta\mathbf{B}}{\delta u}$$

On taking the limits as $\delta u \to 0$, the result follows.

Let **A** be a vector function of the scalar variable u and let u be a scalar function of the scalar variable t. Let $\delta\mathbf{A}$ and δu be the increments of **A** and u which correspond to the increment δt, then

$$\frac{\delta\mathbf{A}}{\delta t} = \frac{\delta\mathbf{A}}{\delta u} \times \frac{\delta u}{\delta t}$$

is an algebraic identity.

On taking the limits as $\delta t \to 0$, we have

$$\frac{d\mathbf{A}}{dt} = \frac{d\mathbf{A}}{du}\frac{du}{dt} \tag{3.1}$$

Let ϕ and **A** be respectively scalar and vector functions of u and let $\delta\phi$ and $\delta\mathbf{A}$ be their increments which correspond to the increment δu. Therefore

$$\begin{aligned}\delta(\phi\mathbf{A}) &= (\phi + \delta\phi)(\mathbf{A} + \delta\mathbf{A}) - \phi\mathbf{A} \\ &= \delta\phi\mathbf{A} + \phi\delta\mathbf{A} + \delta\phi\,\delta\mathbf{A}\end{aligned}$$

i.e.

$$\frac{\delta(\phi\mathbf{A})}{\delta u} = \frac{\delta\phi}{\delta u}\mathbf{A} + \phi\frac{\delta\mathbf{A}}{\delta u} + \frac{\delta\phi}{\delta u}\delta\mathbf{A}$$

In the limit as $\delta u \to 0$,

$$\frac{d(\phi\mathbf{A})}{du} = \frac{d\phi}{du}\mathbf{A} + \phi\frac{d\mathbf{A}}{du} \tag{3.2}$$

Writing **A** in terms of its rectangular components, we have

$$\mathbf{A}(u) = A_1(u)\mathbf{i} + A_2(u)\mathbf{j} + A_3(u)\mathbf{k}$$

where **i**, **j**, **k** are constant unit vectors, so that

$$\frac{d\mathbf{A}}{du} = \frac{dA_1}{du}\mathbf{i} + \frac{dA_2}{du}\mathbf{j} + \frac{dA_3}{du}\mathbf{k} \tag{3.3}$$

3.3 Curves in Space

Each value of u leads to a single value of $\mathbf{A}$. Let $\mathbf{A}$ be the position vector of P relative to an origin O. As u varies continuously so does $\mathbf{A}(u)$, and P traces out a continuous curve in space (Fig. 3.2).

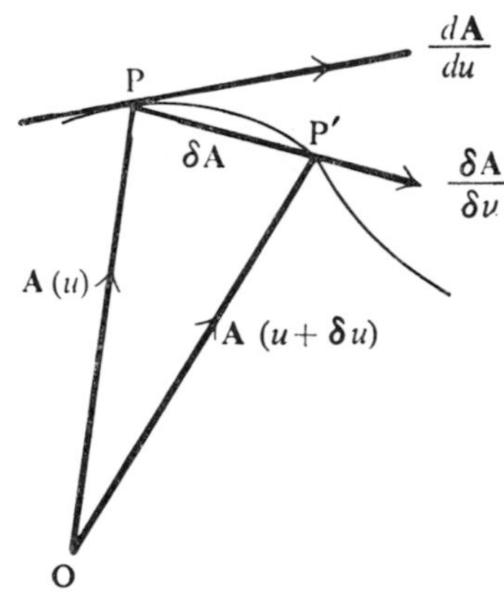

Figure 3.2

Let P, P′ be the positions of the moving point corresponding to u and $u + \delta u$ respectively. Then

$$\mathbf{OP} = \mathbf{A}(u) \quad \text{and} \quad \mathbf{OP}' = \mathbf{A}(u + \delta u)$$

Therefore

$$\delta\mathbf{A} = \mathbf{PP}' = \mathbf{A}(u + \delta u) - \mathbf{A}(u)$$

so that $\delta\mathbf{A}/\delta u$ is a vector parallel to $\mathbf{PP}'$.

In the limit as $\delta u \to 0$ the chord PP′ of the space curve tends to coincidence with the tangent to the curve at P. Thus

$$\frac{d\mathbf{A}}{du} = \lim_{\delta u \to 0} \frac{\delta\mathbf{A}}{\delta u}$$

is a vector having the direction of the tangent at P in the sense of increasing u.

We may write

$$\mathbf{A}(u) = A_1(u)\mathbf{i} + A_2(u)\mathbf{j} + A_3(u)\mathbf{k}$$

so that the space curve has the parametric equations

$$x = A_1(u) \qquad y = A_2(u) \qquad z = A_3(u)$$

and

$$\frac{d\mathbf{A}}{du} = \frac{dA_1}{du}\mathbf{i} + \frac{dA_2}{du}\mathbf{j} + \frac{dA_3}{du}\mathbf{k}$$

Using the customary notation, let $\mathbf{r}(u)$ be the position vector of the poin P(x, y, z) so that

$$\mathbf{r}(u) = x(u)\mathbf{i} + y(u)\mathbf{j} + z(u)\mathbf{k}$$

If now the parameter u represents the time t, $d\mathbf{r}/dt$ will represent the vecto velocity $\mathbf{v}$ of the point P in its path and is given by

$$\mathbf{v} = \frac{d\mathbf{r}}{dt} = \frac{dx}{dt}\mathbf{i} + \frac{dy}{dt}\mathbf{j} + \frac{dz}{dt}\mathbf{k}$$

and $\mathbf{v}$ has a direction which is always tangential to the path of P.

Again if $\mathbf{A}$ represents the velocity vector $\mathbf{v}$, $d\mathbf{v}/dt$ will be the acceleratio vector $\mathbf{a}$.

In kinematics, the abbreviations $\dot{\mathbf{r}}$, $\ddot{\mathbf{r}}$, . . . are widely used to denote firs second, . . . derivatives of $\mathbf{r}$ with respect to the time t. Thus $\mathbf{v} = \dot{\mathbf{r}}$, $\mathbf{a} = \dot{\mathbf{v}} = \ddot{\mathbf{r}}$, and so on.

EXAMPLE 3.1

(A) A particle moves along a curve whose parametric equations are

$$x = a \sin pt \qquad y = a \cos pt \qquad z = bt$$

where t is the time $(t > 0)$ and a, b, p are constants. Determine the velocit and acceleration vectors at time t and show that their magnitudes are constan

Solution The position vector of the particle is

$$\mathbf{r} = (a \sin pt)\mathbf{i} + (a \cos pt)\mathbf{j} + (bt)\mathbf{k}$$

$$\mathbf{v} = \frac{d\mathbf{r}}{dt} = (ap \cos pt)\mathbf{i} + (-ap \sin pt)\mathbf{j} + \mathbf{b}\mathbf{k}$$

$$\mathbf{a} = \frac{d^2\mathbf{r}}{dt^2} = (-ap^2 \sin pt)\mathbf{i} + (-ap^2 \cos pt)\mathbf{j} + 0\mathbf{k}$$

Therefore

$$|\mathbf{v}| = \sqrt{(a^2p^2\cos^2 pt + a^2p^2\sin^2 pt + b^2)} = \sqrt{(a^2p^2 + b^2)}$$

$$|\mathbf{a}| = ap^2$$

(B) With reference to (A), find the components of the velocity and acceleration at time $t = 0$ in the direction of the vector $3\mathbf{i} + 2\mathbf{j} - 6\mathbf{k}$.

Solution At time $t = 0$, $\mathbf{v} = ap\mathbf{i} + b\mathbf{k}$

$$\mathbf{a} = -ap^2\mathbf{j}$$

A unit vector in the direction of the vector $3\mathbf{i} + 2\mathbf{j} - 6\mathbf{k}$ is

$$\frac{3\mathbf{i} + 2\mathbf{j} - 6\mathbf{k}}{7} = \mathbf{d} \quad \text{say}$$

The components of $\mathbf{v}$ and $\mathbf{a}$ in this direction are respectively

$$\mathbf{v.d} = \tfrac{3}{7}(ap - 2b) \quad \text{and} \quad \mathbf{a.d} = -\tfrac{2}{7}ap^2$$

EXERCISES 3.1

1. Given $\mathbf{A}(u) = (a\cos u)\mathbf{i} + (a\sin u)\mathbf{j} + (au\cot\alpha)\mathbf{k}$, show that

$$\left|\frac{d\mathbf{A}}{du}\right| = a\,\text{cosec}\,\alpha \quad \text{and} \quad \left|\frac{d^2\mathbf{A}}{du^2}\right| = a$$

2. A particle moves along a curve whose parametric equations are

$$x = 3t \qquad y = 2t^2 \qquad z = t - t^3$$

where t is the time. Show that the components of its velocity and acceleration in the direction of the vector $2\mathbf{i} - \mathbf{j} - \mathbf{k}$ at $t = 2$ are respectively $9/\sqrt{6}$ and $8/\sqrt{6}$.

3. A particle of mass m is moving in a plane under the influence of a force $\mathbf{F}$ so that its cartesian coordinates in the plane at time t are $[3a\sin(t/t_0), 3a\cos(t/t_0)]$ where a is a constant length and t_0 is a constant time. Describe the motion of the particle as fully as you can and find the time which elapses before the particle returns next to its starting point at $t = 0$.

Express the components of the particle's acceleration in the directions of the two axes and hence prove that $\mathbf{F}$ is of constant magnitude.

Describe carefully how the direction of $\mathbf{F}$ varies with time. At time $t = 4\pi t_0$, the force $\mathbf{F}$ is suddenly annihilated. Find the position of the particle at $t = 8\pi t_0$.

Describe a physical situation of which this question could be a mathematical model. (*Oxford and Cambridge A-level, S.M.P*)

[Uniform motion in a circle radius $3a$; angular velocity $1/t_0$; initial position $\mathbf{r} = 3a\mathbf{j}$; returns after time $2\pi t_0$; $\mathbf{F}$ of constant magnitude $3ma/t_0^2$; when $t = 8\pi t_0$, $\mathbf{r} = 12\pi a\mathbf{i} + 3a\mathbf{j}$]

3.4 Derivatives of Scalar and Vector Products

The method employed to find the derivatives of the products of algebraic functions is directly applicable to the various products of vectors. It will be sufficient to use this method to prove that

$$\frac{d}{du}(\mathbf{A} \times \mathbf{B}) = \frac{d\mathbf{A}}{du} \times \mathbf{B} + \mathbf{A} \times \frac{d\mathbf{B}}{du} \tag{3.4}$$

Let $\delta\mathbf{A}$ and $\delta\mathbf{B}$ be the increments in $\mathbf{A}$ and $\mathbf{B}$ corresponding to the increment δu then

$$\begin{aligned}\frac{\delta(\mathbf{A} \times \mathbf{B})}{\delta u} &= \frac{(\mathbf{A} + \delta\mathbf{A}) \times (\mathbf{B} + \delta\mathbf{B}) - \mathbf{A} \times \mathbf{B}}{\delta u} \\ &= \frac{\delta\mathbf{A} \times \mathbf{B} + \mathbf{A} \times \delta\mathbf{B} + \delta\mathbf{A} \times \delta\mathbf{B}}{\delta u} \\ &= \frac{\delta\mathbf{A}}{\delta u} \times \mathbf{B} + \mathbf{A} \times \frac{\delta\mathbf{B}}{\delta u} + \frac{\delta\mathbf{A}}{\delta u} \times \delta\mathbf{B}\end{aligned}$$

Now taking the limit as $\delta u \to 0$, the result follows. In eqn (3.4) the order of the terms in the products is, of course, significant.

In a similar manner, it can readily be shown that

$$\frac{d}{du}(\mathbf{A}.\mathbf{B}) = \frac{d\mathbf{A}}{du}.\mathbf{B} + \mathbf{A}.\frac{d\mathbf{B}}{du} \tag{3.5}$$

from which it follows that

$$\frac{d}{du}(\mathbf{A}.\mathbf{A}) = 2\mathbf{A}.\frac{d\mathbf{A}}{du}$$

Also

$$\frac{d}{du}(\mathbf{A}.\mathbf{A}) = \frac{d}{du}(A^2) = 2A\frac{dA}{du}$$

Therefore

$$\mathbf{A}.\frac{d\mathbf{A}}{du} = A\,\frac{dA}{du} \tag{3.6}$$

If now **A** is a vector of constant magnitude and variable direction, i.e. A is constant,

$$\mathbf{A}.\frac{d\mathbf{A}}{du} = 0 \quad \text{and} \quad \frac{d\mathbf{A}}{du} \text{ is perpendicular to } \mathbf{A}$$

Scalar and vector triple products may be readily differentiated by applying eqns (3.4) and (3.5). The following formulae are obtained

$$\frac{d}{du}\,[\mathbf{A}\;\mathbf{B}\;\mathbf{C}] = \left[\frac{d\mathbf{A}}{du}\,\mathbf{B}\mathbf{C}\right] + \left[\mathbf{A}\,\frac{d\mathbf{B}}{du}\,\mathbf{C}\right] + \left[\mathbf{A}\mathbf{B}\,\frac{d\mathbf{C}}{du}\right] \tag{3.7}$$

$$\frac{d}{du}\,[\mathbf{A}\times(\mathbf{B}\times\mathbf{C})] = \frac{d\mathbf{A}}{du}\times(\mathbf{B}\times\mathbf{C}) + \mathbf{A}\times\left(\frac{d\mathbf{B}}{du}\times\mathbf{C}\right) + \mathbf{A}\times\left(\mathbf{B}\times\frac{d\mathbf{C}}{du}\right) \tag{3.8}$$

In eqn (3.7) the cyclic order must be preserved and in eqn (3.8) the order of the terms in the products is significant.

EXAMPLE 3.2

(A) If **a** is a constant vector and **r** is a vector function of the scalar variable t, find the derivative of $r^3\mathbf{r} + \mathbf{a}\times d\mathbf{r}/dt$ with respect to t.

Solution

$$\frac{d}{dt}\left(r^3\mathbf{r} + \mathbf{a}\times\frac{d\mathbf{r}}{dt}\right) = \frac{d}{dt}\,(r^3\mathbf{r}) + \frac{d}{dt}\left(\mathbf{a}\times\frac{d\mathbf{r}}{dt}\right)$$

$$= r^3\,\frac{d\mathbf{r}}{dt} + 3r^2\,\frac{dr}{dt}\,\mathbf{r} + \mathbf{a}\times\frac{d^2\mathbf{r}}{dt^2}$$

i.e.

$$\frac{d}{dt}\,(r^3\mathbf{r} + \mathbf{a}\times\dot{\mathbf{r}}) = r^3\dot{\mathbf{r}} + 3r^2\dot{r}\mathbf{r} + \mathbf{a}\times\ddot{\mathbf{r}}$$

(B) Show that $\frac{d}{dt}[\mathbf{r}\,\dot{\mathbf{r}}\,\ddot{\mathbf{r}}] = [\mathbf{r}\,\dot{\mathbf{r}}\,\dddot{\mathbf{r}}]$

$$\frac{d}{dt}[\mathbf{r}\,\dot{\mathbf{r}}\,\ddot{\mathbf{r}}] = [\dot{\mathbf{r}}\,\dot{\mathbf{r}}\,\ddot{\mathbf{r}}] + [\mathbf{r}\,\ddot{\mathbf{r}}\,\ddot{\mathbf{r}}] + [\mathbf{r}\,\dot{\mathbf{r}}\,\dddot{\mathbf{r}}]$$

$$= [\mathbf{r}\,\dot{\mathbf{r}}\,\dddot{\mathbf{r}}]$$

since the first and second scalar triple products on the right-hand side are each zero.

(C) If $\mathbf{r} = \mathbf{a}\cos nt + \mathbf{b}\sin nt$, when $\mathbf{a}$, $\mathbf{b}$, n are constants, show that

(i) $\ddot{\mathbf{r}} + n^2\mathbf{r} = 0$ (ii) $\mathbf{r} \times \dot{\mathbf{r}} = n\mathbf{a} \times \mathbf{b}$

(i) $\dot{\mathbf{r}} = \mathbf{a}(-n \sin nt) + \mathbf{b}(n \cos nt)$

$$\ddot{\mathbf{r}} = \mathbf{a}(-n^2 \cos nt) + \mathbf{b}(-n^2 \sin nt) = -n^2\mathbf{r}$$

$$\ddot{\mathbf{r}} + n^2\mathbf{r} = 0$$

(ii) $\mathbf{r} \times \dot{\mathbf{r}} = (\mathbf{a}\cos nt + \mathbf{b}\sin nt) \times n(-\mathbf{a}\sin nt + \mathbf{b}\cos nt)$

$$= n(\cos^2 nt + \sin^2 nt)(\mathbf{a} \times \mathbf{b})$$

$$= n\mathbf{a} \times \mathbf{b}$$

EXAMPLE 3.3

Find the radial and transverse components of the velocity and acceleration of a particle moving on a curve.

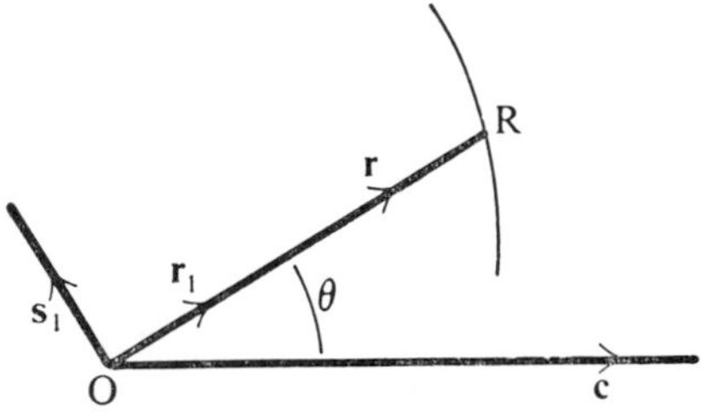

Figure 3.3

Solution Let the position of a point R on the curve (Fig. 3.3) be specified by its polar coordinates (r, θ), θ being measured from the direction of a fixed vector $\mathbf{c}$. Let $\mathbf{r}_1$, $\mathbf{s}_1$ be *unit* vectors along and perpendicular to OR respectively moving with OR.

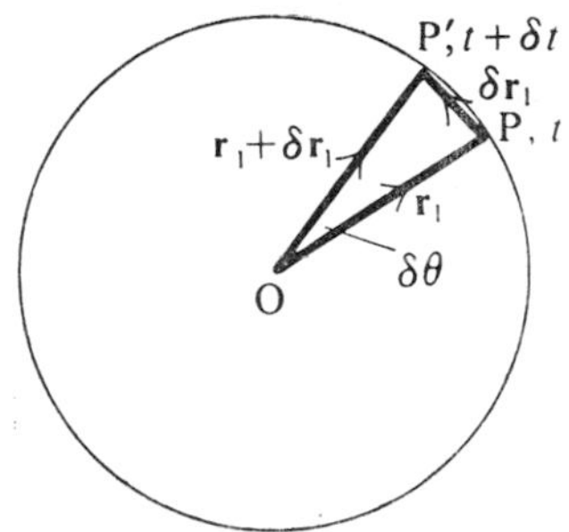

Figure 3.4

As the time t varies, the point P, whose position vector is $\mathbf{r}_1$, moves in a circle of unit radius (Fig. 3.4). If P, P′ are successive positions of this point at times t, $t + \delta t$, then arc PP′ $= \delta\theta$.

Now $\delta\mathbf{r}_1 = \mathbf{PP'}$ and, in the limit, as $\delta\theta \to 0$, $\mathbf{PP'}$ has the direction of the tangent to the circle at P, i.e. has the direction of $\mathbf{s}_1$, and $|\mathbf{PP'}|$ has the magnitude $\delta\theta$. Therefore

$$\frac{d\mathbf{r}_1}{d\theta} = \lim_{\delta\theta \to 0} \frac{\delta\mathbf{r}_1}{\delta\theta} = \mathbf{s}_1$$

$$\frac{d\mathbf{r}_1}{dt} = \frac{d\mathbf{r}_1}{d\theta}\frac{d\theta}{dt} = \dot{\theta}\mathbf{s}_1$$

Similarly

$$\frac{d\mathbf{s}_1}{dt} = -\,\dot{\theta}\mathbf{r}_1$$

Now the velocity $\mathbf{v}$ of the particle at R (Fig. 3.3) is

$$\mathbf{v} = \frac{d\mathbf{r}}{dt} = \frac{d}{dt}(r\mathbf{r}_1)$$

$$= \dot{r}\mathbf{r}_1 + \frac{rd\mathbf{r}_1}{dt} = \dot{r}\mathbf{r}_1 + r\dot{\theta}\mathbf{s}_1$$

Thus the radial and transverse components of the velocity are $\dot{r}$ and $\mathbf{r}\dot{\theta}$ respectively.

Now

$$\mathbf{a} = \frac{d\mathbf{v}}{dt} = \frac{d}{dt}\{\dot{r}\mathbf{r}_1 + r\dot{\theta}\mathbf{s}_1\}$$

$$= \ddot{r}\mathbf{r}_1 + \dot{r}\frac{d\mathbf{r}_1}{dt} + (\dot{r}\dot{\theta} + r\ddot{\theta})\mathbf{s}_1 + r\dot{\theta}\frac{d\mathbf{s}_1}{dt}$$

$$= \ddot{r}\mathbf{r}_1 + \dot{r}\dot{\theta}\mathbf{s}_1 + (\dot{r}\dot{\theta} + r\ddot{\theta})\mathbf{s}_1 - r\dot{\theta}^2\mathbf{r}_1$$

$$= (\ddot{r} - r\dot{\theta}^2)\mathbf{r}_1 + (2\dot{r}\dot{\theta} + r\ddot{\theta})\mathbf{s}_1$$

The radial and transverse components of the acceleration are respectively

$$(\ddot{r} - r\dot{\theta}^2) \quad \text{and} \quad (2\dot{r}\dot{\theta} + r\ddot{\theta}) \quad \text{or} \quad \frac{1}{r}\frac{d}{dt}(r^2\dot{\theta})$$

EXAMPLE 3.4
A point P has coordinates (x, y) with respect to axes Ox, Oy which are rotating about O in their own plane with uniform angular velocity ω. Find expressions for the velocity and acceleration of P.

Solution The position vector of P is

$$\mathbf{r} = x\mathbf{i} + y\mathbf{j}$$

where $\mathbf{i}, \mathbf{j}$ are unit vectors rotating with angular velocity ω. Therefore

$$\mathbf{v} = \dot{\mathbf{r}} = \dot{x}\mathbf{i} + x\frac{d\mathbf{i}}{dt} + \dot{y}\mathbf{j} + y\frac{d\mathbf{j}}{dt}$$

$$= (\dot{x} - \omega y)\mathbf{i} + (\dot{y} + \omega x)\mathbf{j}$$

since $d\mathbf{i}/dt = \omega\mathbf{j}$ and $d\mathbf{j}/dt = -\omega\mathbf{i}$ by Example 3.3.

$$\mathbf{a} = \ddot{\mathbf{r}} = (\ddot{x} - \omega\dot{y})\mathbf{i} + (\dot{x} - \omega y)\omega\mathbf{j} + (\ddot{y} + \omega\dot{x})\mathbf{j} + (\dot{y} + \omega x)(-\omega\mathbf{i})$$

$$= (\ddot{x} - 2\omega\dot{y} - \omega^2 x)\mathbf{i} + (\ddot{y} + 2\omega\dot{x} - \omega^2 y)\mathbf{j}$$

Clearly $(\ddot{x}\mathbf{i} + \ddot{y}\mathbf{j})$ is the acceleration of P relative to the rotating axes. Note that this is the total acceleration of P if the axes were stationary $(\omega = 0)$.

The terms $-\omega^2(x\mathbf{i} + y\mathbf{j}) = -\omega^2\mathbf{r}$ comprise the centripetal acceleration whilst the terms $2\omega(-\dot{y}\mathbf{i} + \dot{x}\mathbf{j})$ are the so-called Coriolis accelerations which are zero when $\omega = 0$.

EXERCISES 3.2

1. If $\mathbf{A} = 6u^2\mathbf{i} - u^3\mathbf{j} + u\mathbf{k}$ and $\mathbf{B} = 2\cos u\,\mathbf{i} - 2\sin u\,\mathbf{j} + \mathbf{k}$, find

   (i) $\dfrac{d}{du}(\mathbf{A.B})$ (ii) $\dfrac{d}{du}(\mathbf{A} \times \mathbf{B})$ (iii) $\dfrac{d}{du}(\mathbf{A.A})$

   [(i) $2u(12 + u^2)\cos u - 6u^2 \sin u + 1$ (ii) $\{2(\sin u + u\cos u) - 3u^2\}\mathbf{i}$ $+ \{2(\cos u - u\sin u) - 12u\}\mathbf{j} - 2\{(12u + u^3)\sin u + 3u^2\cos u\}\mathbf{k}$
   (iii) $2u(1 + 72u^2 + 3u^4)$]

2. Show that

   (*a*) $\dfrac{d}{dt}(\tfrac{1}{2}m\dot{r}^2) = m\dot{\mathbf{r}}.\ddot{\mathbf{r}}$ (*b*) $\dfrac{d}{dt}(\mathbf{r} \times \dot{\mathbf{r}}) = \mathbf{r} \times \ddot{\mathbf{r}}$

   (*c*) $\dfrac{d}{dt}\{\mathbf{r} \times (\dot{\mathbf{r}} \times \ddot{\mathbf{r}})\} = \dot{\mathbf{r}} \times (\dot{\mathbf{r}} \times \ddot{\mathbf{r}}) + \mathbf{r} \times (\dot{\mathbf{r}} \times \dddot{\mathbf{r}})$

   (*d*) $\dfrac{d}{du}\left(\mathbf{A} \times \dfrac{d\mathbf{B}}{du} - \dfrac{d\mathbf{A}}{du} \times \mathbf{B}\right) = \mathbf{A} \times \dfrac{d^2\mathbf{B}}{du^2} - \dfrac{d^2\mathbf{A}}{du^2} \times \mathbf{B}$

3. A particle moves on the circumference of a circle of radius c with a constant angular velocity ω. Show that the position vector of the particle may be expressed in the form

   $$\mathbf{r} = c\cos\omega t\,\mathbf{i} + c\sin\omega t\,\mathbf{j}$$

   Show that its acceleration $\mathbf{a} = -\omega^2\mathbf{r}$ and that its velocity $\mathbf{v}$ is such that $\mathbf{r.v} = 0$ and $\mathbf{r} \times \mathbf{v} = c^2\omega\mathbf{k}$. Interpret these results.

4. With reference to Example 3.3, show that

   $$\frac{d^2\mathbf{r}_1}{dt^2} = \mathbf{s}_1\ddot{\theta} - \mathbf{r}_1\dot{\theta}^2 \quad\text{and}\quad \frac{d^2\mathbf{s}_1}{dt^2} = -\mathbf{r}_1\ddot{\theta} - \mathbf{s}_1\dot{\theta}^2$$

5. An insect is crawling at a constant velocity v along a spoke of a bicycle wheel towards the circumference whilst the wheel is rotating with constant angular velocity ω. Find the magnitude and direction of the velocity and acceleration of the insect when it is at a distance c from the centre.
   [$\sqrt{(v^2 + \omega^2c^2)}$ at $\tan^{-1}(\omega c/v)$ to the spoke; $\omega\sqrt{(4v^2 + \omega^2c^2)}$ at $\tan^{-1}(-2v/\omega c)$ to the spoke]

3.5 *The Elements of Differential Geometry*

We have seen that the position vector function $\mathbf{r}(u)$, where u is a scalar, defines a curve in space. If the scalar u is taken to be s, the length of the arc

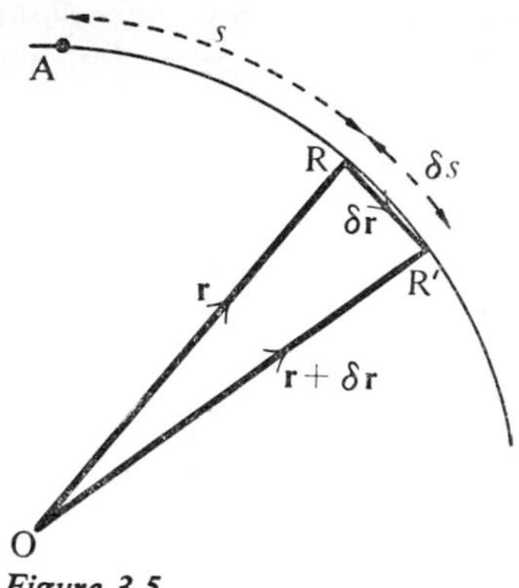

Figure 3.5

measured from any fixed point A on the curve to a variable point R (Fig. 3.5), then the position vector of R is a vector function $\mathbf{r}(s)$ of the scalar s.

Let R, R′ be the points $\mathbf{r}$, $\mathbf{r} + \delta\mathbf{r}$ corresponding to s, $s + \delta s$ respectively. The chord $\mathbf{RR}' = \delta\mathbf{r}$ and the arc $\mathrm{RR}' = \delta s$. As $\delta s \to 0$, the ratio (chord RR′/arc RR′) tends to unity. It follows that

$$\left|\frac{d\mathbf{r}}{ds}\right| = \lim_{\delta s \to 0}\left|\frac{\delta\mathbf{r}}{\delta s}\right| = \lim_{\delta s \to 0}\frac{\text{chord RR}'}{\text{arc RR}'} = 1$$

Also since $\delta\mathbf{r}/\delta s$ is a vector parallel to $\delta\mathbf{r}$, in the limit as $\delta s \to 0$, its direction will be that of the tangent to the curve at R. Therefore

$$\frac{d\mathbf{r}}{ds} = \lim_{\delta s \to 0}\frac{\delta\mathbf{r}}{\delta s} = \mathbf{T} \tag{3.9}$$

where $\mathbf{T}$ is a unit vector having the direction of the tangent to the curve at R in the sense of s increasing.

Now $\mathbf{r}(s) = x(s)\mathbf{i} + y(s)\mathbf{j} + z(s)\mathbf{k}$, so that

$$\frac{d\mathbf{r}}{ds} = \mathbf{T} = \frac{dx}{ds}\mathbf{i} + \frac{dy}{ds}\mathbf{j} + \frac{dz}{ds}\mathbf{k}$$

Thus $(dx/ds, dy/ds, dz/ds)$ are the direction cosines of the *unit tangent* $\mathbf{T}$.

Since $\mathbf{T}.\mathbf{T} = 1$, then $\mathbf{T}.d\mathbf{T}/ds = 0$, and therefore

$$\frac{d\mathbf{T}}{ds} \text{ is perpendicular to } \mathbf{T}$$

Let **N** be a unit vector in the direction of $d\mathbf{T}/ds$, then

$$\frac{d\mathbf{T}}{ds} = \kappa\mathbf{N} \tag{3.10}$$

N, which is perpendicular to **T**, is called the *principal normal*, κ the *curvature*, and $\rho = 1/\kappa$ the *radius of curvature* of the curve at R (Fig. 3.6).

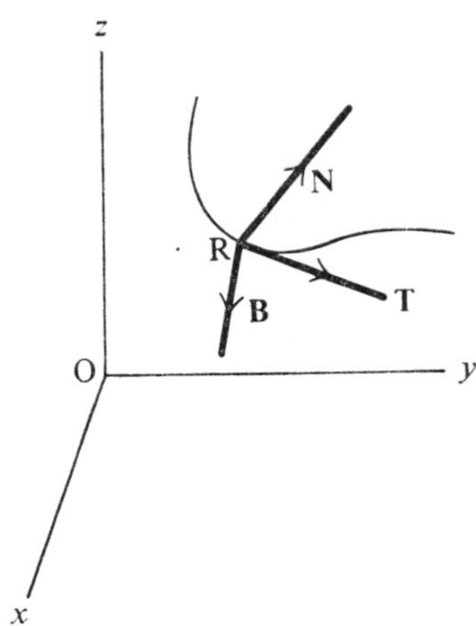

Figure 3.6

Let **B** be a unit vector such that $\mathbf{B} = \mathbf{T} \times \mathbf{N}$. Then **T**, **N**, **B** form a right-handed system of unit vectors.

$$\frac{d\mathbf{B}}{ds} = \frac{d\mathbf{T}}{ds} \times \mathbf{N} + \mathbf{T} \times \frac{d\mathbf{N}}{ds}$$

$$= \kappa\mathbf{N} \times \mathbf{N} + \mathbf{T} \times \frac{d\mathbf{N}}{ds} = \mathbf{T} \times \frac{d\mathbf{N}}{ds}$$

Therefore

$$\mathbf{T}.\frac{d\mathbf{B}}{ds} = \mathbf{T}.\mathbf{T} \times \frac{d\mathbf{N}}{ds} = 0$$

i.e.

$$\frac{d\mathbf{B}}{ds} \text{ is perpendicular to } \mathbf{T}$$

But **B**.**B** = 1, therefore

$$\mathbf{B}.\frac{d\mathbf{B}}{ds} = 0.$$

Thus $d\mathbf{B}/ds$ is perpendicular to both **B** and **T** and is therefore parallel to **N**. Therefore

$$\frac{d\mathbf{B}}{ds} = -\tau\mathbf{N} \tag{3.11}$$

B is called the *binormal*, τ the *torsion* and $\sigma = 1/\tau$ the *radius of torsion* of the curve at R.

Since $\mathbf{N} = \mathbf{B} \times \mathbf{T}$

$$\begin{aligned}\frac{d\mathbf{N}}{ds} &= \frac{d\mathbf{B}}{ds} \times \mathbf{T} + \mathbf{B} \times \frac{d\mathbf{T}}{ds}\\ &= -\tau\mathbf{N} \times \mathbf{T} + \mathbf{B} \times \kappa\mathbf{N}\\ &= \tau\mathbf{B} - \kappa\mathbf{T}\end{aligned} \tag{3.12}$$

The formulae given in eqns (3.10), (3.11) and (3.12):

$$\frac{d\mathbf{T}}{ds} = \kappa\mathbf{N} \qquad \frac{d\mathbf{B}}{ds} = -\tau\mathbf{N} \qquad \frac{d\mathbf{N}}{ds} = \tau\mathbf{B} - \kappa\mathbf{T} \tag{3.13}$$

are known as the Frenet–Serret formulae.

The *osculating plane* to the curve at the point R is the plane containing the tangent and the principal normal. The *normal plane* and the *rectifying plane* are planes through R which are normal to the tangent and the principal normal respectively.

3.6 Equations of Tangent, Normal and Binormal

Let $\mathbf{r}_1$ be the position vector of the point R_1 on a curve and let $\mathbf{T}_1$, $\mathbf{N}_1$, $\mathbf{B}_1$ be the unit tangent, normal and binormal at R_1 respectively (Fig. 3.7). Let **r** be the position vector of any variable point R. If R is on the tangent at R_1 then $(\mathbf{r} - \mathbf{r}_1)$ is parallel to $\mathbf{T}_1$. Therefore

$$(\mathbf{r} - \mathbf{r}_1) \times \mathbf{T}_1 = 0 \tag{3.14}$$

which is the equation of the tangent at R_1.

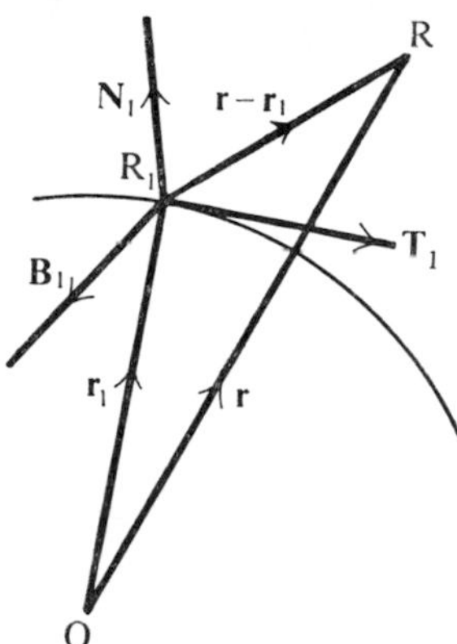

Figure 3..7

Similarly, the equations of the normal and the binormal are respectively

$$(\mathbf{r}-\mathbf{r}_1)\times\mathbf{N}_1=0 \qquad (\mathbf{r}-\mathbf{r}_1)\times\mathbf{B}_1=0$$

If R is in the osculating plane, then $\mathbf{B}_1$ is normal to $(\mathbf{r}-\mathbf{r}_1)$. Thus

$$(\mathbf{r}-\mathbf{r}_1).\mathbf{B}_1=0 \tag{3.15}$$

which is the equation of the osculating plane at R_1.

Similarly, the equations of the normal and rectifying planes are respectively

$$(\mathbf{r}-\mathbf{r}_1).\mathbf{T}_1=0 \qquad (\mathbf{r}-\mathbf{r}_1).\mathbf{N}_1=0$$

EXAMPLE 3.5

(A) Show that

(i) $\kappa=\left|\dfrac{d^2\mathbf{r}}{ds^2}\right|$

(ii) $\dfrac{d\mathbf{r}}{ds}\times\dfrac{d^2\mathbf{r}}{ds^2}=\kappa\mathbf{B}$

(iii) $\left[\dfrac{d\mathbf{r}}{ds}\,\dfrac{d^2\mathbf{r}}{ds^2}\,\dfrac{d^3\mathbf{r}}{ds^3}\right]=\kappa^2\tau$

Solution

$$\text{(i)}\ \frac{d\mathbf{r}}{ds} = \mathbf{T}$$

Thus

$$\frac{d^2\mathbf{r}}{ds^2} = \frac{d\mathbf{T}}{ds} = \kappa\mathbf{N}$$

i.e.

$$\left|\frac{d^2\mathbf{r}}{ds^2}\right| = \kappa$$

$$\text{(ii)}\ \frac{d\mathbf{r}}{ds} \times \frac{d^2\mathbf{r}}{ds^2} = \mathbf{T} \times \kappa\mathbf{N} = \kappa\mathbf{B}$$

$$\text{(iii)}\ \frac{d^3\mathbf{r}}{ds^3} = \frac{d}{ds}(\kappa\mathbf{N}) = \frac{d\kappa}{ds}\mathbf{N} + \kappa\frac{d\mathbf{N}}{ds}$$

$$= \frac{d\kappa}{ds}\mathbf{N} + \kappa(\tau\mathbf{B} - \kappa\mathbf{T})$$

Thus

$$\left[\frac{d\mathbf{r}}{ds}, \frac{d^2\mathbf{r}}{ds^2}, \frac{d^3\mathbf{r}}{ds^3}\right] = \kappa\mathbf{B}.\left\{\frac{d\kappa}{ds}\mathbf{N} + \kappa\tau\mathbf{B} - \kappa^2\mathbf{T}\right\} = \kappa^2\tau \text{ by (ii)}$$

(B) Show for the plane curve $y = f(x)$, $z = 0$, that $\rho = (1 + y'^2)^{3/2}/y''$ where a dash indicates differentiation with respect to x.

Solution $\mathbf{r} = x\mathbf{i} + y\mathbf{j}$. Therefore

$$\mathbf{T} = \frac{d\mathbf{r}}{ds} = \frac{d\mathbf{r}}{dx}\bigg/\frac{ds}{dx} = \frac{\mathbf{i} + y'\mathbf{j}}{(1 + y'^2)^{1/2}}$$

since $ds/dx = |d\mathbf{r}/dx|$. Therefore, if $\mathbf{N}$ is the unit vector,

$$\frac{d\mathbf{T}}{ds} = \frac{\mathbf{N}}{\rho} = \frac{d\mathbf{T}}{dx}\bigg/\frac{ds}{dx}$$

$$= \frac{(1 + y'^2)^{1/2}y''\mathbf{j} - (\mathbf{i} + y'\mathbf{j})[y'y''/(1 + y'^2)^{1/2}]}{(1 + y'^2)} \cdot (1 + y'^2)^{-1/2}$$

$$= \frac{y''}{(1 + y'^2)^{3/2}} \cdot \frac{(-y'\mathbf{i} + \mathbf{j})}{(1 + y'^2)^{1/2}} = \frac{y''}{(1 + y'^2)^{3/2}}\mathbf{N}$$

where $\mathbf{N}$ is the unit vector $(-y'\,\mathbf{i} + \mathbf{j})/(1 + y'^2)^{1/2}$

Therefore $\rho = \dfrac{(1 + y'^2)^{3/2}}{y''}$

EXAMPLE 3.6

A curve is defined by the parametric equations

$$x = \tan^{-1} s \qquad y = (1/\sqrt{2}) \log (1 + s^2) \qquad z = s - \tan^{-1} s$$

where the parameter s measures the length of the curve. Find τ and κ and the equations of the tangent and of the osculating plane at the point $s = 1$.

$$\mathbf{r} = (\tan^{-1} s)\mathbf{i} + \left\{\frac{1}{\sqrt{2}} \log (1 + s^2)\right\}\mathbf{j} + (s - \tan^{-1} s)\mathbf{k}$$

Therefore

$$\mathbf{T} = \frac{d\mathbf{r}}{ds} = \frac{1}{(1 + s^2)}\{\mathbf{i} + s\sqrt{2}\mathbf{j} + s^2\mathbf{k}\}$$

$$\frac{d\mathbf{T}}{ds} = \kappa\mathbf{N} = \frac{(1 + s^2)\{\sqrt{2}\mathbf{j} + 2s\mathbf{k}\} - 2s\{\mathbf{i} + s\sqrt{2}\mathbf{j} + s^2\mathbf{k}\}}{(1 + s^2)^2}$$

$$= \frac{\sqrt{2}}{(1 + s^2)} \frac{\{-\sqrt{2}s\mathbf{i} + (1 - s^2)\mathbf{j} + \sqrt{2}s\mathbf{k}\}}{(1 + s^2)}$$

$$= \frac{\sqrt{2}}{(1 + s^2)}\mathbf{N}$$

Therefore

$$\kappa = \frac{\sqrt{2}}{(1 + s^2)} \quad \text{and} \quad \mathbf{N} = \frac{1}{(1 + s^2)} \{-\sqrt{2}s\mathbf{i} + (1 - s^2)\mathbf{j} + \sqrt{2}s\mathbf{k}\}$$

Now

$$\mathbf{B} = \mathbf{T} \times \mathbf{N} = \frac{1}{(1 + s^2)^2} \begin{vmatrix} \mathbf{i} & \mathbf{j} & \mathbf{k} \\ 1 & \sqrt{2}s & s^2 \\ -\sqrt{2}s & (1 - s^2) & \sqrt{2}s \end{vmatrix}$$

$$= \frac{1}{(1 + s^2)^2} \{s^2(1 + s^2)\mathbf{i} - s\sqrt{2}(1 + s^2)\mathbf{j} + (1 + s^2)\mathbf{k}\}$$

$$= \frac{1}{(1 + s^2)} \{s^2\mathbf{i} - s\sqrt{2}\mathbf{j} + \mathbf{k}\}$$

$$\frac{d\mathbf{B}}{ds} = -\tau\mathbf{N} = \frac{2s\mathbf{i} + \sqrt{2}(s^2 - 1)\mathbf{j} - 2s\mathbf{k}}{(1 + s^2)^2}$$

$$= \frac{-\sqrt{2}}{(1 + s^2)} \left\{\frac{-s\sqrt{2}\mathbf{i} + (1 - s^2)\mathbf{j} + s\sqrt{2}\mathbf{k}}{(1 + s^2)}\right\}$$

$$\frac{-\sqrt{2}}{(1 + s^2)} \mathbf{N}$$

Therefore

$$\tau = \frac{\sqrt{2}}{(1 + s^2)} = \kappa$$

At the point $s = 1$, i.e. $x_1 = \tan^{-1} 1 = \frac{1}{4}\pi$, $y_1 = (1/\sqrt{2}) \log 2$, $z_1 = 1 - \frac{1}{4}\pi$,

$$\tau = \kappa = 1/\sqrt{2} \qquad \mathbf{T}_1 = \tfrac{1}{2}(\mathbf{i} + \sqrt{2}\,\mathbf{j} + \mathbf{k})$$

$$\mathbf{N}_1 = \tfrac{1}{2}(-\sqrt{2}\,\mathbf{i} + \sqrt{2}\,\mathbf{k}) \qquad \mathbf{B}_1 = \tfrac{1}{2}(\mathbf{i} - \sqrt{2}\,\mathbf{j} + \mathbf{k})$$

The equation of the tangent at $\mathbf{r}_1$ on the curve is given by

$$(\mathbf{r} - \mathbf{r}_1) \times \mathbf{T}_1 = 0$$

or by

$$(\mathbf{r} - \mathbf{r}_1) = t\mathbf{T}_1$$

where t is a parameter. In either case,

$$\frac{x - \frac{1}{4}\pi}{1} = \frac{y - (1/\sqrt{2}) \log 2}{\sqrt{2}} = \frac{z - 1 + \frac{1}{4}\pi}{1}$$

The equation of the osculating plane at $\mathbf{r}_1$ is given by

$$(\mathbf{r} - \mathbf{r}_1).\mathbf{B}_1 = 0 \quad \text{or} \quad [(\mathbf{r} - \mathbf{r}_1), \mathbf{T}_1, \mathbf{N}_1] = 0$$

In either case,

$$(x - \tfrac{1}{4}\pi).1 + \left(y - \frac{1}{\sqrt{2}} \log 2\right).(-\sqrt{2}) + (z - 1 + \tfrac{1}{4}\pi).1 = 0$$

i.e. $x - y\sqrt{2} + z = 1 - \log 2$.

EXAMPLE 3.7

The *circular helix* is a curve lying on the surface of a cylinder cutting its generators at a constant angle. Taking the axis of the cylinder as the z-axis, the parametric equations of the helix will be

$$x = a \cos t \qquad y = a \sin t \qquad z = at \cot \alpha$$

where a is the radius of the cylinder.

$$\mathbf{r} = (a \cos t)\mathbf{i} + (a \sin t)\mathbf{j} + (at \cot \alpha)\mathbf{k}$$

Now

$$\mathbf{T} = \frac{d\mathbf{r}}{ds} = \frac{d\mathbf{r}}{dt}\frac{dt}{ds} = \{(-a \sin t)\mathbf{i} + (a \cos t)\mathbf{j} + (a \cot \alpha)\mathbf{k}\} \Big/ \frac{ds}{dt}$$

But

$$\frac{ds}{dt} = \left|\frac{d\mathbf{r}}{dt}\right| = \sqrt{\{a^2(\sin^2 t + \cos^2 t) + a^2 \cot^2 \alpha\}} = a \operatorname{cosec} \alpha$$

Therefore

$$\mathbf{T} = \{(-\sin t)\mathbf{i} + (\cos t)\mathbf{j} + (\cot \alpha)\mathbf{k}\} \sin \alpha$$

Now $\mathbf{k}$ is a unit vector parallel to the generators, and therefore the angle of intersection ϕ of the curve with a generator through the point t is given by

$$\cos \phi = \mathbf{T}.\mathbf{k} = \cos \alpha$$

Therefore $\phi = \alpha =$ constant. Thus the helix cuts the generators at a constant angle α.

$$\frac{d\mathbf{T}}{ds} = \kappa\mathbf{N} = \{(-\cos t)\mathbf{i} - (\sin t)\mathbf{j}\} \sin \alpha \frac{\sin \alpha}{a}$$

Therefore

$$\kappa = \left|\frac{d\mathbf{T}}{ds}\right| = \frac{\sin^2 \alpha}{a} \quad \text{and} \quad \mathbf{N} = -\{(\cos t)\mathbf{i} + (\sin t)\mathbf{j}\}$$

Therefore $\mathbf{N}.\mathbf{k} = 0$, so that the principal normal is perpendicular to the axis of the cylinder and intersects it.

$$\mathbf{B} = \mathbf{T} \times \mathbf{N} = -\sin \alpha \begin{vmatrix} \mathbf{i} & \mathbf{j} & \mathbf{k} \\ -\sin t & \cos t & \cot \alpha \\ \cos t & \sin t & 0 \end{vmatrix}$$

$$= -\sin \alpha\{(-\cot \alpha \sin t)\mathbf{i} + (\cot \alpha \cos t)\mathbf{j} - \mathbf{k}\}$$

Now $\mathbf{B}.\mathbf{k} = \sin \alpha$, and therefore the binormals are inclined at a constant angle $(\frac{1}{2}\pi - \alpha)$ to the generators.

$$\frac{d\mathbf{B}}{ds} = -\tau\mathbf{N} = \sin \alpha\{(\cot \alpha \cos t)\mathbf{i} + (\cot \alpha \sin t)\mathbf{j}\} \frac{\sin \alpha}{a}$$

$$= -\frac{\sin \alpha \cos \alpha}{a} \mathbf{N}$$

Therefore

$$\tau = \frac{1}{a}\sin\alpha\cos\alpha$$

EXAMPLE 3.8

Find the tangential and normal components of the acceleration of a particle moving on a curve.

Solution Let t be time, and s the arc distance. Then

$$\mathbf{v} = \frac{d\mathbf{r}}{dt} \quad \text{and} \quad v = \frac{ds}{dt} = \left|\frac{d\mathbf{r}}{dt}\right|$$

Therefore $\mathbf{v} = v\mathbf{T}$ and

$$\mathbf{a} = \frac{d\mathbf{v}}{dt} = \frac{d}{dt}(v\mathbf{T}) = \frac{dv}{dt}\mathbf{T} + v\frac{d\mathbf{T}}{dt}$$

$$= \frac{dv}{dt}\mathbf{T} + v\frac{d\mathbf{T}}{ds}\frac{ds}{dt}$$

$$= \frac{dv}{dt}\mathbf{T} + \frac{v^2}{\rho}\mathbf{N}$$

Thus the tangential and normal components of $\mathbf{a}$ are dv/dt and v^2/ρ respectively, and $\mathbf{a}$ lies in the osculating plane.

EXAMPLE 3.9

The equation of a curve in space is given by $\mathbf{r} = \mathbf{r}(t)$. Show that

$$\kappa = \frac{|\dot{\mathbf{r}} \times \ddot{\mathbf{r}}|}{|\dot{\mathbf{r}}|^3} \quad \text{and} \quad \tau = \frac{[\dot{\mathbf{r}}, \ddot{\mathbf{r}}, \dddot{\mathbf{r}}]}{|\dot{\mathbf{r}} \times \ddot{\mathbf{r}}|^2}$$

Solution

$$\frac{d\mathbf{r}}{dt} = \frac{d\mathbf{r}}{ds}\frac{ds}{dt} \qquad \text{i.e. } \dot{\mathbf{r}} = \dot{s}\mathbf{T}$$

and $\dot{s} = |\dot{\mathbf{r}}|$ if s is measured in the direction of increasing t. Therefore

$$\ddot{\mathbf{r}} = \ddot{s}\mathbf{T} + \dot{s}\frac{d\mathbf{T}}{ds}\dot{s} = \ddot{s}\mathbf{T} + \kappa\dot{s}^2\mathbf{N}$$

Therefore $\dot{\mathbf{r}} \times \ddot{\mathbf{r}} = \kappa\dot{s}^3\mathbf{B}$ i.e. $\kappa = |\dot{\mathbf{r}} \times \ddot{\mathbf{r}}|/|\dot{\mathbf{r}}|^3$

Now

$$\dddot{\mathbf{r}} = \dddot{s}\mathbf{T} + \kappa\dot{s}\ddot{s}\mathbf{N} + \frac{d}{dt}(\kappa\dot{s}^2)\mathbf{N} + \kappa\dot{s}^3(\tau\mathbf{B} - \kappa\mathbf{T})$$

Therefore

$$[\dot{\mathbf{r}}, \ddot{\mathbf{r}}, \dddot{\mathbf{r}}] = (\dot{\mathbf{r}} \times \ddot{\mathbf{r}}).\dddot{\mathbf{r}} = (\kappa\dot{s}^3)^2\tau\mathbf{B}.\mathbf{B}$$

$$\tau = \frac{[\dot{\mathbf{r}}, \ddot{\mathbf{r}}, \dddot{\mathbf{r}}]}{|\dot{\mathbf{r}} \times \ddot{\mathbf{r}}|^2}$$

EXERCISES 3.3

1. Show that a curve, whose curvature is everywhere zero, is a straight line.
2. Show that for a plane curve, $\tau = 0$ everywhere.
3. Show that (i) $\dfrac{d\mathbf{T}}{ds} \times \dfrac{d^2\mathbf{T}}{ds^2} = \kappa^2\tau\mathbf{T} + \kappa^3\mathbf{B}$

   (ii) $\left[\dfrac{d\mathbf{T}}{ds}, \dfrac{d^2\mathbf{T}}{ds^2}, \dfrac{d^3\mathbf{T}}{ds^3}\right] = \kappa^5\dfrac{d}{ds}\left(\dfrac{\tau}{\kappa}\right)$
4. Verify that the Frenet–Serret formulae may be written in the form

$$\frac{d\mathbf{T}}{ds} = \mathbf{A} \times \mathbf{T} \qquad \frac{d\mathbf{N}}{ds} = \mathbf{A} \times \mathbf{N} \qquad \frac{d\mathbf{B}}{ds} = \mathbf{A} \times \mathbf{B}$$

   where $\mathbf{A} = \tau\mathbf{T} + \kappa\mathbf{B}$.
5. Find the radius of curvature of the curve whose position vector is given by

$$\mathbf{r} = (a\cos u)\mathbf{i} + (b\sin u)\mathbf{j}$$

   a, b being constants. Comment on this result for $a = b$.

   $[(a^2\sin^2 u + b^2\cos^2 u)^{3/2}/ab]$

6. Show that for the plane curve $x = x(t)$, $y = y(t)$

$$\kappa = \frac{|\dot{x}\ddot{y} - \dot{y}\ddot{x}|}{(\dot{x}^2 + \dot{y}^2)^{3/2}}$$

7. Find the vectors **T**, **N** and **B** for the curve defined by

$$\mathbf{r} = t\mathbf{i} + t^2\mathbf{j} + \tfrac{2}{3}t^3\mathbf{k}$$

and show that $\kappa = \tau = 2/(1 + 2t^2)^2$.

$$[\mathbf{T} = (1 + 2t^2)^{-1}(\mathbf{i} + 2t\mathbf{j} + 2t^2\mathbf{k})$$
$$\mathbf{N} = (1 + 2t^2)^{-1}\{-2t\mathbf{i} + (1 - 2t^2)\mathbf{j} + 2t\mathbf{k}\}$$
$$\mathbf{B} = (1 + 2t^2)^{-1}(2t^2\mathbf{i} - 2t\mathbf{j} + \mathbf{k})]$$

8. The parametric equations of the twisted cubic are $x = t$, $y = t^2$, $z = t^3$. Show that

$$\kappa = \frac{2(9t^4 + 9t^2 + 1)^{1/2}}{(9t^4 + 4t^2 + 1)^{3/2}} \qquad \tau = \frac{3}{9t^4 + 9t^2 + 1}$$

9. Using the results of Example 3.9 show for the curve defined by

$$\mathbf{r} = (3t - t^3)\mathbf{i} + 3t^2\mathbf{j} + (3t + t^3)\mathbf{k}$$

that $\kappa = \tau = 1/3(1 + t^2)^2$.

10. For Exercise 9 show that

$$\mathbf{T} = 2^{-1/2}(1 + t^2)^{-1}\{(1 - t^2)\mathbf{i} + 2t\mathbf{j} + (1 + t^2)\mathbf{k}\}$$
$$\mathbf{N} = -(1 + t^2)^{-1}\{2t\mathbf{i} + (t^2 - 1)\mathbf{j}\}$$
$$\mathbf{B} = 2^{-1/2}(1 + t^2)^{-1}\{(t^2 - 1)\mathbf{i} - 2t\mathbf{j} + (1 + t^2)\mathbf{k}\}$$

Obtain the equations of the osculating, normal and rectifying planes at the point $t = 1$. $[y - z + 1 = 0,\ y + z - 7 = 0,\ x = 2]$

11. Show that for the curve $x = 16\cos^3 t$, $y = 16\sin^3 t$, $z = 9\cos 2t$,

$$\mathbf{T} = \tfrac{1}{5}\{(-4\cos t)\mathbf{i} + (4\sin t)\mathbf{j} - 3\mathbf{k}\}$$
$$\mathbf{N} = (\sin t)\mathbf{i} + (\cos t)\mathbf{j}$$
$$\mathbf{B} = \tfrac{1}{5}\{(3\cos t)\mathbf{i} - (3\sin t)\mathbf{j} - 4\mathbf{k}\}$$

Find the equations of the tangent, normal and binormal and of the osculating, normal and rectifying planes at the point $t = \pi/3$.

$$\left[\text{Tangent: } \frac{x - 2}{2} = \frac{y - 6\sqrt{3}}{-2\sqrt{3}} = \frac{x + 9/2}{3}\right.$$

$$\text{Normal: } \frac{x - 2}{\sqrt{3}} = \frac{y - 6\sqrt{3}}{1} = \frac{z + 9/2}{0}$$

$$\text{Binormal: } \frac{x - 2}{-3} = \frac{y - 6\sqrt{3}}{3\sqrt{3}} = \frac{z + 9/2}{8}$$

Osculating plane: $3x - 3\sqrt{3}y - 8z + 12 = 0$; normal: $4x - 4\sqrt{3}y + 6z + 91 = 0$; rectifying: $\sqrt{3}x + y - 8\sqrt{3} = 0$]

12. Show that the acceleration vector **a** of a particle moving on a curve always lies in the osculating plane.

13. A particle is moving on the curve

$$\mathbf{r} = 3t^3\mathbf{i} + (t - 2t^2)\mathbf{j} - 2t^3\mathbf{k}$$

where t is the time. Find the magnitudes of the tangential and normal components of its acceleration at time $t = 1$. [$41\sqrt{(2/7)}$; $\sqrt{(26/7)}$]

14. A particle, moving on a curve, has instantaneously a velocity **v** and an acceleration **a**. Show that the curvature of its path at this instant is given by

$$\kappa = v^{-3}\,|\mathbf{v} \times \mathbf{a}|$$

4 *Vector integration*

4.1 *Integral of a Vector Function of a Scalar*

If $\mathbf{F}(u)$ is a vector function of a scalar u such that

$$\frac{d\mathbf{F}}{du} = \mathbf{R}(u)$$

then we say that $\mathbf{F}(u)$ is the integral of $\mathbf{R}(u)$ with respect to u, i.e.

$$\mathbf{F}(u) = \int \mathbf{R}(u)\, du$$

as in the case of algebraic functions.

If $\mathbf{c}$ is an arbitrary *constant* vector,

$$\frac{d}{du}(\mathbf{F} + \mathbf{c}) = \frac{d\mathbf{F}}{du} = \mathbf{R}(u)$$

i.e.

$$\int \mathbf{R}(u)\, du = \mathbf{F} + \mathbf{c}$$

$\mathbf{F}(u)$ is called the *indefinite* integral of $\mathbf{R}(u)$ since it is indefinite to the extent of the arbitrary constant of integration $\mathbf{c}$. A *definite* integral, between specified limits $u = a$ and $u = b$, is denoted by

$$\int_b^a \mathbf{R}(u)\, du$$

and may be evaluated in the same way as definite integrals of algebraic functions.

From section 3.4, it follows that

$$\int \left(\frac{d\mathbf{A}}{du}.\mathbf{B} + \mathbf{A}.\frac{d\mathbf{B}}{du}\right) du = \mathbf{A}.\mathbf{B} + c \tag{4.1}$$

and

$$\int \left(\frac{d\mathbf{A}}{du} \times \mathbf{B} + \mathbf{A} \times \frac{d\mathbf{B}}{du}\right) du = \mathbf{A} \times \mathbf{B} + \mathbf{c} \tag{4.2}$$

From (4.1) we have

$$\int \mathbf{A}.\frac{d\mathbf{A}}{du}\, du = \tfrac{1}{2}\mathbf{A}.\mathbf{A} + c = \tfrac{1}{2}A^2 + c$$

and hence

$$2\int \mathbf{r}.\dot{\mathbf{r}}\, dt = (\mathbf{r})^2 + c \tag{4.3}$$

$$2\int \dot{\mathbf{r}}.\ddot{\mathbf{r}}\, dt = (\dot{\mathbf{r}})^2 + c \tag{4.4}$$

From (4.2) we have

$$\int (\dot{\mathbf{r}} \times \dot{\mathbf{r}} + \mathbf{r} \times \ddot{\mathbf{r}})\, dt = \mathbf{r} \times \dot{\mathbf{r}} + \mathbf{c}$$

i.e.

$$\int (\mathbf{r} \times \ddot{\mathbf{r}})\, dt = \mathbf{r} \times \dot{\mathbf{r}} + \mathbf{c} \tag{4.5}$$

EXAMPLE 4.1
If $\mathbf{R}(u) = u^3\mathbf{i} + (1 - u)\mathbf{j} - 3u^2\mathbf{k}$, find $\int \mathbf{R}(u)\, du$ and $\int_1^2 \mathbf{R}(u)\, du$.

$$\int \mathbf{R}(u)\, du = \mathbf{i}\int u^3\, du + \mathbf{j}\int (1 - u)\, du - \mathbf{k}\int 3u^2\, du$$

$$= \tfrac{1}{4}u^4\mathbf{i} + (u - \tfrac{1}{2}u^2)\mathbf{j} - u^3\mathbf{k} + \mathbf{c}$$

$$\int_1^2 \mathbf{R}(u)\, du = [\tfrac{1}{4}u^4\mathbf{i} + (u - \tfrac{1}{2}u^2)\mathbf{j} - u^3\mathbf{k}]_1^2$$

$$= \tfrac{1}{4}(16 - 1)\mathbf{i} + (2 - 1 - 2 + \tfrac{1}{2})\mathbf{j} - (8 - 1)\mathbf{k}$$

$$= \tfrac{1}{4}\{15\mathbf{i} - 2\mathbf{j} - 28\mathbf{k}\}$$

EXAMPLE 4.2
The acceleration $\mathbf{a}$ of a particle at time t is given by

$$\mathbf{a} = (\cos t)\mathbf{i} + \mathrm{e}^{-t}\mathbf{j} - 6t\mathbf{k}$$

If the velocity $\mathbf{v}$ and the displacement $\mathbf{r}$ at time $t = 0$ are $-\mathbf{j}$ and $-\mathbf{i}$ respectively, find $\mathbf{v}$ and $\mathbf{r}$ at time t.

Solution

$$\mathbf{a} = \frac{d\mathbf{v}}{dt} = (\cos t)\mathbf{i} + \mathrm{e}^{-t}\mathbf{j} - 6t\mathbf{k}$$

Therefore

$$\mathbf{v} = \int \{(\cos t)\mathbf{i} + \mathrm{e}^{-t}\mathbf{j} - 6t\mathbf{k}\}\, dt$$

$$= (\sin t)\mathbf{i} - \mathrm{e}^{-t}\mathbf{j} - 3t^2\mathbf{k} + \mathbf{c}$$

When $t = 0$, $\mathbf{v} = -\mathbf{j}$. Therefore

$$-\mathbf{j} = -\mathbf{j} + \mathbf{c} \qquad \mathbf{c} = 0$$

$$\mathbf{v} = \frac{d\mathbf{r}}{dt} = (\sin t)\mathbf{i} - \mathrm{e}^{-t}\mathbf{j} - 3t^2\mathbf{k}$$

$$\mathbf{r} = -(\cos t)\mathbf{i} + \mathrm{e}^{-t}\mathbf{j} - t^3\mathbf{k} + \mathbf{d}$$

When $t = 0$, $\mathbf{r} = -\mathbf{i}$. Therefore

$$-\mathbf{i} = -\mathbf{i} + \mathbf{j} + \mathbf{d} \qquad \mathbf{d} = -\mathbf{j}$$

$$\mathbf{r} = -(\cos t)\mathbf{i} + (\mathrm{e}^{-t} - 1)\mathbf{j} - t^3\mathbf{k}$$

EXAMPLE 4.3
The position vector $\mathbf{r}$ of a moving particle is given by the following equation of motion

$$\ddot{\mathbf{r}} + n^2\mathbf{r} = 0 \qquad \text{(see Example 3.2(C), p. 70)}$$

To integrate this equation, form the scalar product with $\dot{\mathbf{r}}$, and we have

$$2\dot{\mathbf{r}}.\ddot{\mathbf{r}} + 2n^2\mathbf{r}.\dot{\mathbf{r}} = 0$$

On integrating with respect to t, we have

$$(\dot{\mathbf{r}})^2 + n^2(\mathbf{r})^2 = c \quad \text{or} \quad (\dot{\mathbf{r}})^2 = v^2 = c - \mu(\mathbf{r})^2 \quad \text{where} \quad \mu = n^2 \tag{4.6}$$

Again on taking the vector product of the equation of motion with $\mathbf{r}$, we have

$$\mathbf{r} \times \ddot{\mathbf{r}} + \mu\mathbf{r} \times \mathbf{r} = 0 \quad \text{i.e. } \mathbf{r} \times \ddot{\mathbf{r}} = 0$$

On integrating with respect to t, we have

$$\mathbf{r} \times \dot{\mathbf{r}} = \mathbf{h} \quad \text{a constant vector}$$

i.e.

$$\mathbf{r} \times \mathbf{v} = \mathbf{h} = pv\mathbf{k} \tag{4.7}$$

where p is the perpendicular from O to the tangent to the path at R (Fig. 4.1) and $\mathbf{k}$ is a unit vector having the direction of the vector $\mathbf{r} \times \mathbf{v}$. Therefore

$$pv = h \tag{4.8}$$

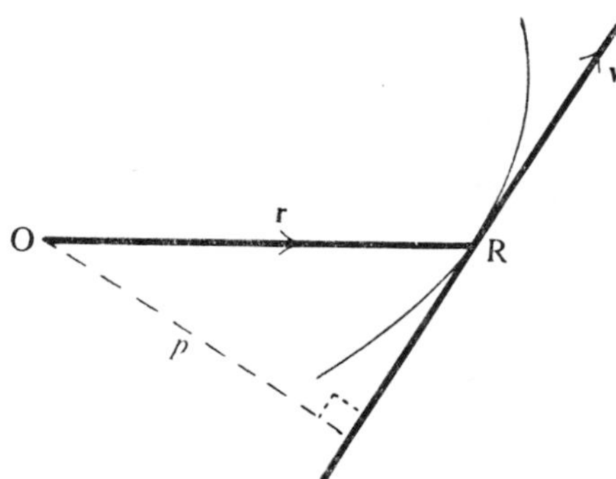

Figure 4.1

Since $\mathbf{h}$ is a constant vector perpendicular to the plane of $\mathbf{r}$ and $\mathbf{v}$, the particle moves in a plane curve. (See section 4.9.)

EXERCISES 4.1

1. If $\mathbf{A}(u) = u^2\mathbf{i} + (u - 1)\mathbf{j} - u\mathbf{k}$ and $\mathbf{B} = -u\mathbf{i} + (u + 1)\mathbf{j} - u^2\mathbf{k}$, find $\int_0^3 \mathbf{A}.\mathbf{B}\, du$ and $\int \mathbf{A} \times \mathbf{B}\, du$. $[6, 3\frac{1}{3}\mathbf{i} + 9\frac{1}{15}\mathbf{j} + 7\frac{1}{3}\mathbf{k}]$
2. If $\mathbf{A}(t) = 2t\mathbf{i} - t^3\mathbf{j} + (1 - t^2)\mathbf{k}$, find $\int_1^2 \mathbf{A}.(d\mathbf{A}/dt)\, dt$ [42]

4.2 Line Integrals

Let C be a curve defined by the position vector

$$\mathbf{r}(u) = x(u)\mathbf{i} + y(u)\mathbf{j} + z(u)\mathbf{k}$$

and let $u = u_1$, $u = u_2$ determine two points P, Q respectively on C.

Let $\mathbf{A}(x, y, z) = A_1\mathbf{i} + A_2\mathbf{j} + A_3\mathbf{k}$ be a vector function of position which

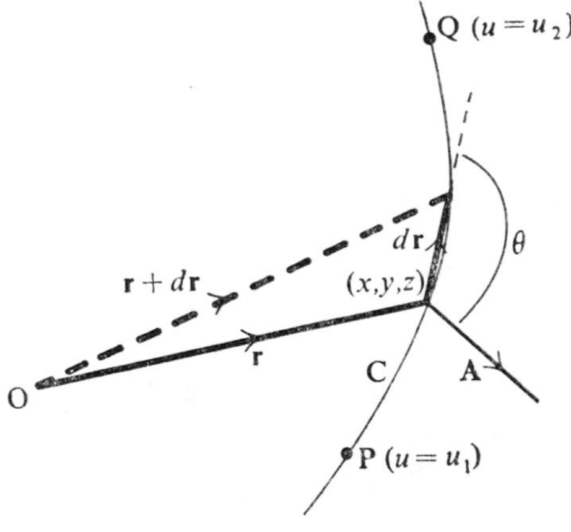

Figure 4.2

is defined and continuous at all points on C. See Fig. 4.2. Now

$$\mathbf{A}.d\mathbf{r} = A \cos\theta\, dr = dr \times \text{the tangential component of } \mathbf{A}$$

The line integral of $\mathbf{A}$ along the curve between P and Q is defined to be the integral of the tangential component of **A** between these points, i.e.

$$\int_{P}^{Q} \mathbf{A}.d\mathbf{r} = \int_{P}^{Q} (A_1\, dx + A_2\, dy + A_3\, dz)$$

$$= \int_{u_2}^{u_2} \left(A_1 \frac{dx}{du} + A_2 \frac{dy}{du} + A_3 \frac{dz}{du}\right) du$$

If a particle is moving along C under the action of a force $\mathbf{F} = X\mathbf{i} + Y\mathbf{j} + Z\mathbf{k}$ then the line integral

$$\int_{P}^{Q} \mathbf{F}.d\mathbf{r} = \int_{P}^{Q} (X\, dx + Y\, dy + Z\, dz)$$

represents the work done by **F** in moving the particle from P to Q. If C is a simple closed curve, i.e. a closed curve which does not intersect itself, the line integral of **A** around C is denoted by

$$\oint \mathbf{A}.d\mathbf{r} = \oint (A_1\, dx + A_2\, dy + A_3\, dz)$$

If the velocity **V** of a fluid is given by $\mathbf{V} = u\mathbf{i} + v\mathbf{j} + w\mathbf{k}$, then in aero- and hydrodynamics

$$\oint \mathbf{V}.d\mathbf{r} = \oint (u\, dx + v\, dy + w\, dz)$$

is termed the *circulation* about the curve C.

EXAMPLE 4.4

A particle moves in a force field $\mathbf{F} = x^2\mathbf{i} + (y^2 - z^2)\mathbf{j} + z^3\mathbf{k}$. Find the work done by **F** in moving the particle along

(*a*) the straight line joining the points (0, 0, 0) and (1, 2, 1)

(*b*) the curve $x = t^3$, $y = 2t^2$, $z = t + 2$ from $t = 0$ to $t = 1$

(*c*) the curve of intersection of the surfaces $2y^2 = x$ and $y^2 = z$ from $y = 0$ to $y = 1$, i.e. from the origin to the point (2, 1, 1).

Solution

(*a*) The equation of the straight line is

$$\frac{x}{1} = \frac{y}{2} = \frac{z}{1} = t \qquad \text{i.e. } x = z = t,\ y = 2t$$

$$\text{Work done} = \int \mathbf{F}.d\mathbf{r} = \int \{x^2\,dx + (y^2 - z^2)\,dy + z^3\,dz\}$$

$$= \int_0^1 \{t^2\,dt + (4t^2 - t^2)\,d(2t) + t^3dt\}$$

$$= \int_0^1 (t^3 + 7t^2)\,dt$$

$$= \left[\frac{t^4}{4} + \frac{7t^3}{3}\right]_0^1 = 2\tfrac{7}{12}$$

(*b*) $$\text{Work done} = \int_0^1 [t^6 3t^2\,dt + \{4t^4 - (t+2)^2\}4t\,dt + (t+2)^3\,dt]$$

$$= \int_0^1 (3t^8 + 16t^5 - 3t^3 - 10t^2 - 4t + 8)\,dt$$

$$= \left[\frac{t^9}{3} + \frac{8t^6}{3} - \frac{3t^4}{4} - \frac{10t^3}{3} - 2t^2 + 8t\right]_0^1 = 4\tfrac{11}{12}$$

(*c*) $$\text{Work done} = \int \{x^2\,dx + (y^2 - z^2)\,dy + z^3\,dz\} \text{ where } x = 2y^2,\ z = y^2$$

$$= \int_0^1 \{4y^4 4y\,dy + (y^2 - y^4)\,dy + y^6 2y\,dy\}$$

$$= \int_0^1 \{16y^5 + y^2 - y^4 + 2y^7\}\,dy$$

$$= \left[\frac{8y^6}{3} + \frac{y^3}{3} - \frac{y^5}{5} + \frac{y^8}{4}\right]_0^1 = 3\tfrac{1}{20}$$

EXAMPLE 4.5

Calculate the circulation of a vector $\mathbf{V} = (x - 2y)\mathbf{i} + (2x + y)\mathbf{j}$ in an anticlockwise sense around the closed curve bounded by the curves $y = x^2$, $8x = y^2$ in the plane $z = \theta$.

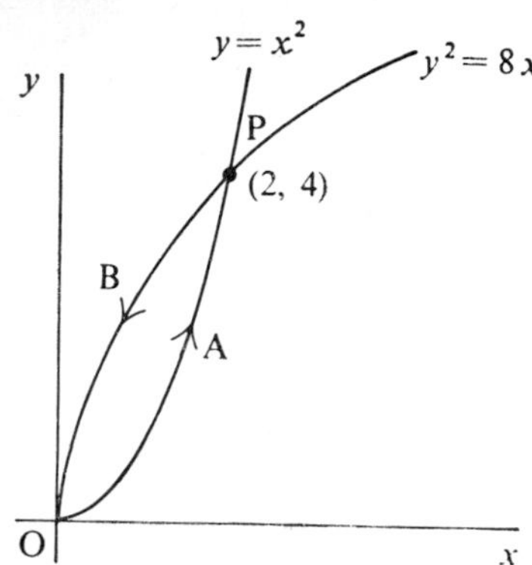

Figure 4.3

Solution These two boundary curves meet at the point (2, 4) (Fig. 4.3).

$$\text{Circulation} = \oint \mathbf{V}.d\mathbf{r}$$

$$= \int \{(x - 2y)\,dx + (2x + y)\,dy\}$$

$$= \int_0^2 \{(x - 2x^2)\,dx + (2x + x^2)2x\,dx\}$$

$$+ \int_4^0 \left\{\left(\frac{y^2}{8} - 2y\right)\frac{2y\,dy}{8} + \left(\frac{y^2}{4} + y\right)dy\right\}$$

since $y = x^2$ along OAP and x goes from 0 to 2 whilst $x = y^2/8$ along PBO and y goes from 4 to 0. Therefore

$$\text{Circulation} = \int_0^2 (x + 2x^2 + 2x^3)\,dx - \int_0^4 \left(\frac{y^3}{32} - \frac{y^2}{4} + y\right)dy$$

$$= \left[\frac{x^2}{2} + \frac{2x^3}{3} + \frac{x^4}{2}\right]_0^2 - \left[\frac{y^4}{128} - \frac{y^3}{12} + \frac{y^2}{2}\right]_0^4 = 10\tfrac{2}{3}$$

EXERCISES 4.2

1. If $\mathbf{A} = (2x - y)\mathbf{i} + (x + z)\mathbf{j} + (y + z)\mathbf{k}$, evaluate $\int \mathbf{A}.d\mathbf{r}$ along the following curves:
   (*a*) straight lines from the origin to (0, 0, 2), thence to (0, 1, 2), and finally to (2, 1, 2)
   (*b*) $x = 2t$, $y = t^3$, $z = 2t^2$ from the origin to the point (2, 1, 2) [(*a*)6, (*b*)9]
2. Find the circulation of the vector

$$\mathbf{A} = (x^2 - y^2)\mathbf{i} + (x^2 + y^2)\mathbf{j}$$

around the triangle whose vertices are (0, 0), (0, 3), (1, 3) in the counterclockwise sense (Fig. 4.4). [7]

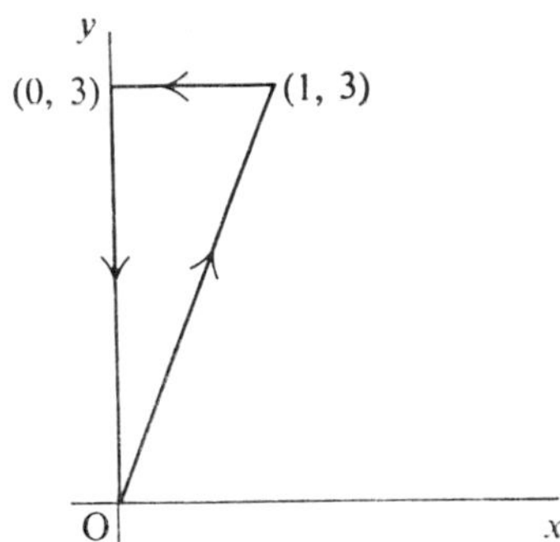

Figure 4.4

3. Show that the line integral of the vector

$$\mathbf{A} = (2xz + y^2)\mathbf{i} + 2xy\mathbf{j} + x^2z\mathbf{k}$$

along a curve between two given points is independent of the curve.
[Hint: show that $(2xz + y^2)\,dx + 2xy\,dy + x^2z\,dz$ is an exact differential]
4. Show that the circulation of the vector $(xy - z)\mathbf{i} + (yz - x)\mathbf{j} + (zx - y)\mathbf{k}$ in the clockwise sense around the circle $x^2 + y^2 = 4$, $z = 2$ is 4π.

4.3 *Surface Integrals*

Let S be a surface having two sides. Either side of the surface may be arbitrarily chosen as the positive side. Let dS be an element of area on the surface and let $\mathbf{n}$ be the unit normal to the surface at dS drawn outwardly from the positive side.

The vector area $d\mathbf{S}$ is taken to be a vector of magnitude dS having the direction of $\mathbf{n}$, i.e. $d\mathbf{S} = dS\,\mathbf{n}$.

Integrals of the type

$$\iint_S \mathbf{A}.d\mathbf{S} = \iint_S \mathbf{A}.\mathbf{n}\,dS$$

evaluated over the surface S arise frequently in applied mathematics.

$\iint \mathbf{A}.d\mathbf{S}$ is called the flux of **A** over S, and is a scalar quantity.

$$\iint_S \mathbf{A} \times d\mathbf{S} = \iint_S \mathbf{A} \times \mathbf{n}\,dS$$

is another example of a surface integral—a vector quantity. Surface integrals may be evaluated in terms of double integrals taken over the area obtained by projecting S on to one of the coordinate planes. In Fig. 4.5, S_{xy} is the

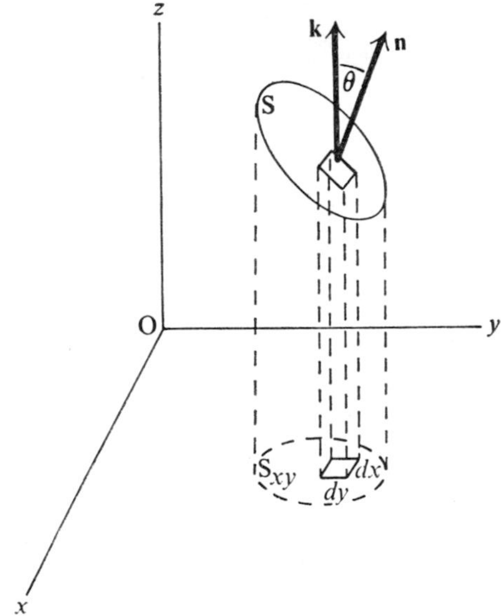

Figure 4.5

projection of S on the x,y plane and the surface integral may be found by integrating over S_{xy} with respect to x,y or other appropriate coordinates.

This procedure is possible provided that perpendiculars to the x,y plane meet S in one point only. If this is not so, S may be divided into sub-areas each of which satisfies this requirement. The surface integral is then found as the sum of the integrals over the sub-areas.

Consider the evaluation of $\iint_S \mathbf{A}.d\mathbf{S} = \iint_S \mathbf{A}.\mathbf{n}\, dS$. Let $dx\, dy$ be the projection of dS on the x,y plane (Fig. 4.5). Since $\mathbf{n}$ is a unit normal to dS and $\mathbf{k}$ is a unit normal to $dx\, dy$,

$$dx\, dy = dS \cos\theta = (\mathbf{n}.\mathbf{k})\, dS \qquad \text{i.e. } dS = \frac{dx\, dy}{(\mathbf{n}.\mathbf{k})}$$

Therefore

$$\iint_S \mathbf{A}.d\mathbf{S} = \iint_{S_{xy}} \mathbf{A}.\mathbf{n} \frac{dx\, dy}{(\mathbf{n}.\mathbf{k})} \tag{4.9}$$

4.4 Normal to a Surface

The vector differential operator del, denoted by ∇, is defined by

$$\nabla \equiv \frac{\partial}{\partial x}\mathbf{i} + \frac{\partial}{\partial y}\mathbf{j} + \frac{\partial}{\partial z}\mathbf{k} \tag{4.10}$$

Let $\phi(x, y, z)$ be a defined and differentiable function in a given region. The *gradient* of ϕ is denoted by grad ϕ or $\nabla\phi$ and is defined by

$$\nabla\phi = \left(\frac{\partial}{\partial x}\mathbf{i} + \frac{\partial}{\partial y}\mathbf{j} + \frac{\partial}{\partial z}\mathbf{k}\right)\phi = \frac{\partial\phi}{\partial x}\mathbf{i} + \frac{\partial\phi}{\partial y}\mathbf{j} + \frac{\partial\phi}{\partial z}\mathbf{k} \tag{4.11}$$

Let $\mathbf{r} = x\mathbf{i} + y\mathbf{j} + z\mathbf{k}$ be the position vector of a point (x, y, z) on the surface $\phi =$ constant, then $d\mathbf{r} = dx\,\mathbf{i} + dy\,\mathbf{j} + dz\,\mathbf{k}$ lies in the tangent plane to the surface at (x, y, z).

Now $d\phi = \phi(x + dx, y + dy, z + dz) - \phi(x, y, z)$ is of the second order of small quantities. Therefore

$$\frac{\partial\phi}{\partial x}dx + \frac{\partial\phi}{\partial y}dy + \frac{\partial\phi}{\partial z}dz = 0$$

i.e. $$\left(\frac{\partial\phi}{\partial x}\mathbf{i} + \frac{\partial\phi}{\partial y}\mathbf{j} + \frac{\partial\phi}{\partial z}\mathbf{k}\right).(dx\,\mathbf{i} + dy\,\mathbf{j} + dz\,\mathbf{k}) = 0$$

so that $\nabla\phi.d\mathbf{r} = 0$. Thus $\nabla\phi$ is a vector perpendicular to $d\mathbf{r}$ and therefore to the surface. Therefore $\nabla\phi$ is a vector having the direction of the unit normal $\mathbf{n}$ to $\phi(x, y, z) =$ constant at (x, y, z).

EXAMPLE 4.6

Evaluate the surface integral $\iint_S \mathbf{A}.\mathbf{n}\, dS$ of the vector $\mathbf{A} = y\mathbf{i} + 2z\mathbf{j} - x\mathbf{k}$ over that part of the plane $2x + y + 3z = 6$ which lies in the first octant.

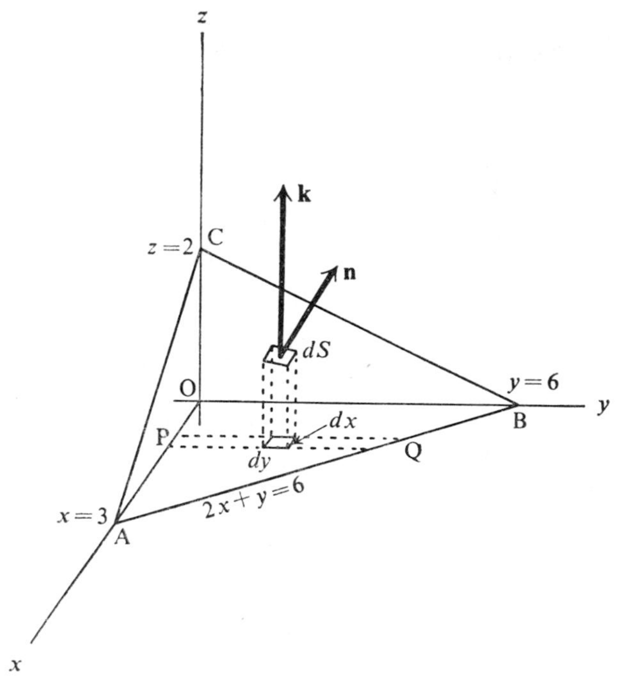

Figure 4.6

Solution The surface integral is to be evaluated over the triangular area ABC (Fig. 4.6). By projecting on to the x,y plane, the surface integral may be evaluated as

$$\iint_{\Delta OAB} \mathbf{A}.\mathbf{n} \frac{dx\, dy}{(\mathbf{n}.\mathbf{k})}$$

when **n** is a unit vector normal to plane ABC. Therefore

$$\mathbf{n} = \frac{2\mathbf{i} + \mathbf{j} + 3\mathbf{k}}{\sqrt{14}} \qquad \mathbf{n}.\mathbf{k} = \frac{3}{\sqrt{14}}$$

$$\mathbf{A}.\mathbf{n} = \frac{1}{\sqrt{14}}(2y + 2z - 3x)$$

Therefore

$$\text{Surface integral} = \iint_{\Delta OAB} \frac{1}{\sqrt{14}}(2y + 2z - 3x)\frac{dx\,dy}{3/\sqrt{14}}$$

$$= \frac{1}{3}\iint_{\Delta OAB} (2y + 2z - 3x)\,dx\,dy$$

But on the plane, $2z = 4 - 2y/3 - 4x/3$, and therefore

$$\text{Surface integral} = \frac{1}{3}\iint_{\Delta OAB}\left(\frac{4y}{3} - \frac{13x}{3} + 4\right)dx\,dy$$

It remains to determine the limits of integration for x and y. The equations of the line AB are $2x + y = 6$, $z = 0$. Keep x constant and integrate along PQ, i.e. from $y = 0$ to $y = 6 - 2x$. The whole of the area OAB will now be covered if we integrate from $x = 0$ to $x = 3$.

$$\text{Surface integral} = \frac{1}{9}\int_{x=0}^{3}\int_{y=0}^{6-2x}(4y - 13x + 12)\,dx\,dy$$

$$= \frac{1}{9}\int_{x=0}^{3}[2y^2 - 13xy + 12y]_0^{6-2x}\,dx$$

$$= \frac{1}{9}\int_{0}^{3}(144 - 150x + 34x^2)\,dx = 7$$

EXAMPLE 4.7

Evaluate $\iint_S \mathbf{A}.d\mathbf{S}$ where $\mathbf{A} = x\mathbf{i} - y\mathbf{j} + 2z\mathbf{k}$ over that part of the surface of the parabolic cylinder $4y = x^2$ cut off by the planes $y = 4$ and $z = 5$.

Solution $\phi = x^2 - 4y = 0$

$$\nabla\phi = \left(\frac{\partial}{\partial x}\mathbf{i} + \frac{\partial}{\partial y}\mathbf{j} + \frac{\partial}{\partial z}\mathbf{k}\right)\phi = 2x\mathbf{i} - 4\mathbf{j} + 0\mathbf{k}$$

$$\mathbf{n} = \frac{x\mathbf{i} - 2\mathbf{j}}{\sqrt{(x^2 + 4)}} \quad \text{and} \quad \mathbf{A.n} = \frac{1}{\sqrt{(x^2 + 4)}}\{x^2 + 2y\}$$

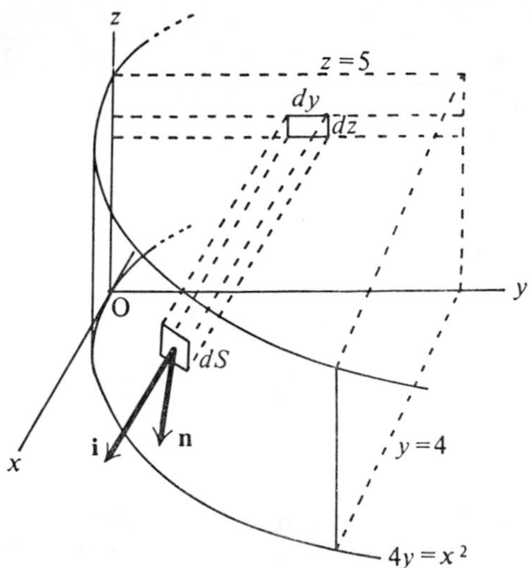

Figure 4.7

In this example it is clearly not possible to evaluate the surface integral by projection on the x,y plane. However

$$\iint_S \mathbf{A}.d\mathbf{S} = \iint_{S_{yz}} \mathbf{A}.\mathbf{n}\,\frac{dy\,dz}{(\mathbf{n}.\mathbf{i})}$$

by projecting on the y, z plane and $\mathbf{n}.\mathbf{i} = x/\sqrt{(x^2+4)}$. Therefore

$$\text{Surface integral} = \iint_{S_{yz}} \frac{1}{\sqrt{(x^2+4)}}(x^2+2y^2)\frac{dy\,dz}{x/\sqrt{(x^2+4)}}$$

$$= \iint_{S_{yz}} \frac{x^2+2y}{x}\,dy\,dz$$

But on the surface $x^2 = 4y$ and therefore

$$\text{Surface intergal} = \int_{z=0}^{5}\int_{y=0}^{4} \frac{4y+2y}{2y^{1/2}}\,dy\,dz$$

$$= \int_{z=0}^{5}\int_{y=0}^{4} 3y^{1/2}dy\,dz$$

$$= \int_0^5 [2y^{3/2}]_0^4\,dz$$

$$= \int_0^5 16\,dz = 80$$

EXERCISES 4.3

1. Evaluate $\iint_S \mathbf{A}.d\mathbf{S}$ when $\mathbf{A} = (x^2 + y)\mathbf{i} - 2y\mathbf{j} + xz\mathbf{k}$ and S is that part of the plane $x + 2y + 2z = 10$ which is in the first octant. [250]
2. If $\mathbf{r} = x\mathbf{i} + y\mathbf{j} + z\mathbf{k}$, show that $\iint \mathbf{r}.d\mathbf{S}$ evaluated over all the faces of a unit cube bounded by the planes $x = 0$, $x = 1$, $y = 0$, $y = 1$, $z = 0$, $z = 1$ is 3.
3. If $\mathbf{A} = 2yz\mathbf{i} + xz\mathbf{j} + z\mathbf{k}$, evaluate $\iint \mathbf{A}.d\mathbf{S}$ over the closed surface bounded by the right cone $x^2 + y^2 = 4z^2$ and the plane $z = 2$. [$160\pi/3$]
4. The equation of a surface S is $\phi(x, y, z) = 0$. Show that its area is given by

$$\iint_{S_{xy}} \left\{\left(\frac{\partial\phi}{\partial x}\right)^2 + \left(\frac{\partial\phi}{\partial y}\right)^2 + \left(\frac{\partial\phi}{\partial z}\right)^2\right\}^{1/2} \left\{\left|\frac{\partial\phi}{\partial z}\right|\right\}^{-1} dx\,dy$$

   Use this result to find the area of the surface of the sphere $x^2 + y^2 + z^2 = a^2$.
5. Find the area of the surface of the plane $3x + y + 2z = 6$ cut off by (*a*) the planes $x = 0$, $y = 0$, $x = 1$, $y = 2$ and (*b*) the planes $x = 0$, $y = 0$ and the cylinder $4x^2 + y^2 = 4$. [(*a*) $\sqrt{14}$, (*b*) $\pi\sqrt{14}/4$]
6. Find the surface area of the region common to the two cylinders

$$x^2 + y^2 = 4 \quad \text{and} \quad x^2 + z^2 = 4. \quad [64]$$

7. A hole of square cross-section, side $2b$, is cut symmetrically through a right circular cylinder of radius $a > b$, the axis of the hole being perpendicular to, and two faces of the hole being parallel to, the axis of the cylinder.
   Show that the area removed from the curved surface of the cylinder is $8ab \times \sin^{-1}(b/a)$.
8. Show that the area of that portion of the surface of the sphere $x^2 + y^2 + z^2 = a^2$ within the cylinder $x^2 + y^2 = ax$ is $2a^2(\pi - 2)$.
   [Hint: use cylindrical coordinates]

4.5 *Volume Integrals*

Let V be the volume within a closed surface. Integrals of the types

$$\iiint_V \mathbf{A}\,dV \quad \text{and} \quad \iiint_V \phi\,dV$$

which are integrated over the volume V, with $\mathbf{A}$ and ϕ being respectively vector and scalar functions of position, are called *volume integrals.*

EXAMPLE 4.8
A tetrahedron is bounded by the planes $z = 0$, $z = x$, $y = 2a$, $y = 2x$. If the tetrahedron is composed of material of density $\rho(x, y, z) = xy^2z^3$, calculate its mass.

Solution Mass $= \iiint_{OABC} \rho \, dx \, dy \, dz = \iiint_{OABC} xy^2z^3 \, dx \, dy \, dz$.

ΔABC is right-angled at A (Fig. 4.8). Let $\alpha\beta\gamma$ be a slice of the tetrahedron of width dy parallel to ΔABC.

Consider an element of mass $\rho \, dx \, dy \, dz$ at the point P(x, y, z). Keeping x and y constant, integrate with respect to z so that z goes from 0 to x (the equation of $\beta\gamma$ is $y =$ constant, $z = x$). Then integrate with respect to x so

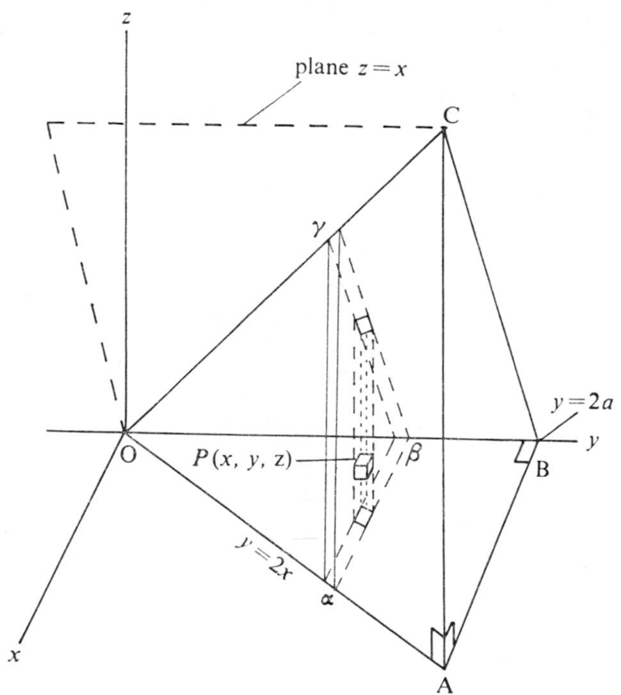

Figure 4.8

that x goes from 0 to $y/2$, and finally integrate with respect to y which goes from 0 to $2a$. Therefore

$$\text{Mass} = \int_{y=0}^{2a} \int_{x=0}^{y/2} \int_{z=0}^{x} xy^2z^3 \, dx \, dy \, dz = \int_{y=0}^{2a} \int_{x=0}^{y/2} xy^2 \left[\frac{z^4}{4}\right]_0^x dx \, dy$$

$$= \frac{1}{4} \int_{y=0}^{2a} \int_{x=0}^{y/2} x^5y^2 \, dx \, dy = \frac{1}{24} \int_0^{2a} \frac{1}{2^6} y^8 \, dy$$

$$= \frac{1}{24} \times \frac{1}{2^6} \times 2^9 \times \frac{a^9}{9} = \frac{a^9}{27}$$

EXAMPLE 4.9
Find the volume common to the sphere $x^2 + y^2 + z^2 = a^2$ and the cylinder $x^2 + y^2 = ax$.

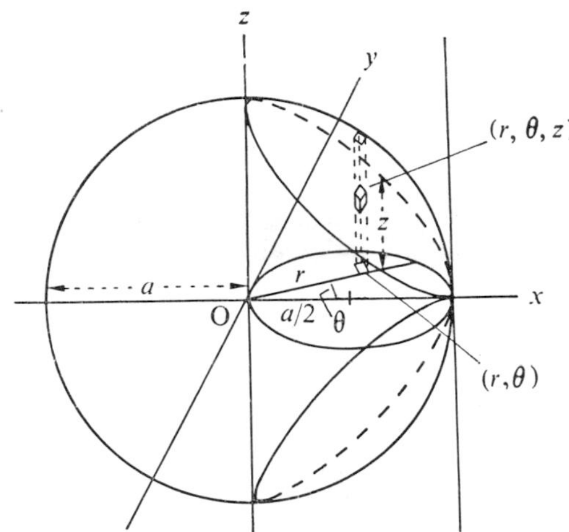

Figure 4.9

Solution The axis of the cylinder is parallel to Oz and the cylinder intersects the plane $z = 0$ in the circle $(x - \frac{1}{2}a)^2 + y^2 = (\frac{1}{2}a)^2$, i.e. a circle centre $(\frac{1}{2}a, 0)$, radius $\frac{1}{2}a$. Using cylindrical coordinates, the element of volume at the point (r, θ, z) is $r\,d\theta\,dr\,dz$.

The volume common to the sphere and the cylinder is $V = \iiint r\,d\theta\,dr\,dz$.
Keeping r, θ constant, integrate with respect to z from $-\sqrt{(a^2 - r^2)}$ to $+\sqrt{(a^2 - r^2)}$.
Now keeping θ constant, integrate with respect to r from 0 to a $\cos\theta$ and finally integrate with respect to θ from $-\frac{1}{2}\pi$ to $+\frac{1}{2}\pi$. Thus

$$V = \int_{\theta=-\pi/2}^{+\pi/2} \int_{r=0}^{a\cos\theta} \int_{z=-\sqrt{(a^2-r^2)}}^{+\sqrt{(a^2-r^2)}} r\,d\theta\,dr\,dz$$

$$= 4\int_{\theta=0}^{\pi/2} \int_{r=0}^{a\cos\theta} r\sqrt{(a^2 - r^2)}\,dr\,d\theta$$

$$= 4\int_{0}^{\pi/2} [-\tfrac{1}{3}(a^2 - r^2)^{3/2}]_0^{a\cos\theta}\,d\theta$$

$$= \frac{4a^3}{3}\int_0^{\pi/2} (1 - \sin^3\theta)\,d\theta = \frac{4a^3}{3}\left(\frac{\pi}{2} - \frac{2}{3}\right) = \frac{2a^3}{9}(3\pi - 4)$$

EXAMPLE 4.10
Evaluate $\iiint_V \mathbf{A}\,dV$ where $\mathbf{A} = x\mathbf{i} + \mathbf{j} - 2y\mathbf{k}$ and where V is the region bounded by the planes $x = 0$, $y = 0$, $z = 0$ and $2x + 2y + z = 4$.

Solution

$$\iiint_V \mathbf{A}\,dV = \mathbf{i}\iiint_V x\,dx\,dy\,dz + \mathbf{j}\iiint_V dx\,dy\,dz - 2\mathbf{k}\iiint_V y\,dx\,dy\,dz$$

Consider the volume element $dx\,dy\,dz$ at the point (x, y, z) (Fig. 4.10). Integrate with respect to z from $z = 0$ to $z = 2(2 - x - y)$. Then integrate with respect to y from $y = 0$ to $y = 2 - x$, and finally integrate with respect to x from $x = 0$ to $x = 2$.

$$\iiint_V dx\,dy\,dz = \int_{x=0}^{2}\int_{y=0}^{2-x}\int_{z=0}^{2(2-x-y)} dx\,dy\,dz$$

$$= 2\int_{x=0}^{2}\int_{y=0}^{2-x}(2 - x - y)\,dx\,dy$$

$$= 2\int_{x=0}^{2}[2y - xy - \tfrac{1}{2}y^2]_0^{2-x}\,dx$$

$$= 2\int_0^2(2 - 2x + \tfrac{1}{2}x^2)\,dx$$

$$= \tfrac{8}{3}$$

$$\iiint_V x\,dx\,dy\,dz = \int_{x=0}^{2}\int_{y=0}^{2-x}\int_{z=0}^{2(2-x-y)} x\,dx\,dy\,dz$$

$$= 2\int_{x=0}^{2}\int_{y=0}^{2-x}x(2 - x - y)\,dx\,dy$$

$$= 2\int_{x=0}^{2}x[2y - xy - \tfrac{1}{2}y^2]_0^{2-x}\,dx$$

$$= 2\int_0^2(2x - 2x^2 + \tfrac{1}{2}x^3)\,dx$$

$$= \tfrac{4}{3}$$

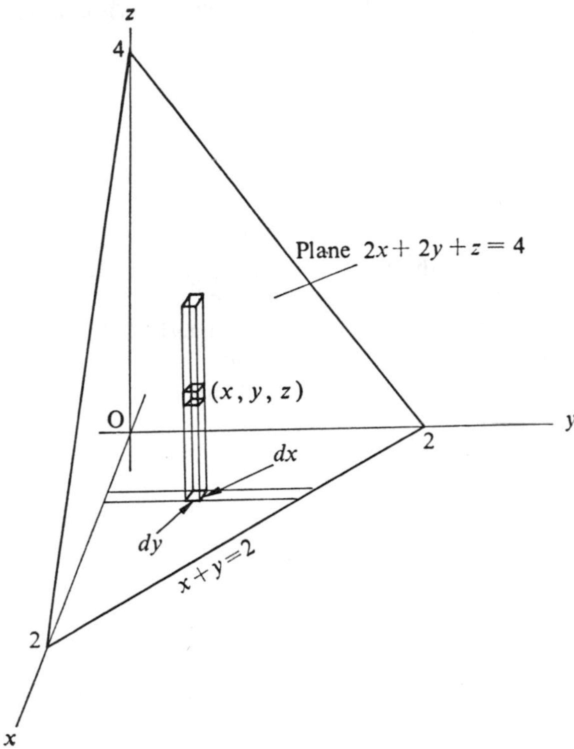

Figure 4.10

$$\iiint_V y\,dx\,dy\,dz = \int_{x=0}^{2}\int_{y=0}^{2-x}\int_{z=0}^{2(2-x-y)} y\,dx\,dy\,dz$$

$$= 2\int_{x=0}^{2}\int_{y=0}^{2-x} y(2-x-y)\,dx\,dy$$

$$= 2\int_{x=0}^{2}[y^2 - \tfrac{1}{2}xy^2 - \tfrac{1}{3}y^3]_0^{2-x}\,dx$$

$$= 2\int_0^2 \tfrac{1}{6}(2-x)^3\,dx = -\frac{1}{3}\left[\frac{(2-x)^4}{4}\right]_0^2$$

$$= \tfrac{4}{3}$$

(The last two integrals must be equal because the figure is symmetrical with respect to x and y.) Therefore

$$\iiint_V \mathbf{A}\, dV = \tfrac{4}{3}(\mathbf{i} + 2\mathbf{j} - 2\mathbf{k})$$

EXERCISES 4.4

1. A tetrahedron is bounded by the planes $x = 0, y = 0, z = 0$ and $3x + 2y + 2z = 6$. Its density $\rho(x, y, z) = xy^2$. Find its mass. [0.9]
2. Evaluate $\iiint z(x^2 + y^2)\, dx\, dy\, dz$ through the volume of the cylinder $x^2 + y^2 = a^2$ intercepted by the planes $z = 0$, $z = h$. [$\frac{1}{4}\pi a^4 h^2$]
3. Evaluate $\iiint (xy + yz + zx)\, dx\, dy\, dz$ through the interior of the cube determined by $0 \leqslant x \leqslant a$, $0 \leqslant y \leqslant a$, $0 \leqslant z \leqslant a$. [$\frac{3}{4}a^5$]
4. Evaluate $\iiint (2x + y)\, dx\, dy\, dz$ over the region bounded by $x = 0$, $y = 0$, $z = 0$, $y = 2$ and the cylinder $z = 4 - x^2$. [$26\frac{2}{3}$]
5. Find the volume bounded by the plane $x + y + 2z = 4a$ and the paraboloid $x^2 + y^2 = 4az$. [$(25/2)\pi a^3$]

Dynamics of a Particle

Let us now consider the applications of vector calculus to dynamics.

4.6 Linear Momentum; Impulse; Activity; Kinetic Energy

The *linear momentum* $\mathbf{p}$ of a particle of mass m moving with a velocity $\mathbf{v}$ is defined by

$$\mathbf{p} = m\mathbf{v} = m\dot{\mathbf{r}} \tag{4.12}$$

By Newton's second law of motion, the force $\mathbf{F}$ acting on the particle is equal to the rate of change of linear momentum which it produces, i.e.

$$\mathbf{F} = \frac{d\mathbf{p}}{dt} = \frac{d}{dt}(m\mathbf{v}) = m\frac{d\mathbf{v}}{dt} = m\mathbf{a} \tag{4.13}$$

where $\mathbf{a}$ is the acceleration of the particle assuming m is constant.

Let the position vector of the particle be

$$\mathbf{r} = x\mathbf{i} + y\mathbf{j} + z\mathbf{k}$$

and let forces $\mathbf{F}_1, \mathbf{F}_2, \ldots, \mathbf{F}_s, \ldots, \mathbf{F}_n$ each act on the particle where $\mathbf{F}_s = X_s\mathbf{i} + Y_s\mathbf{j} + Z_s\mathbf{k}$. The resultant force is

$$\mathbf{R} = \sum_{s=1}^{n} \mathbf{F}_s$$

and the equation of motion is $\mathbf{R} = m\ddot{\mathbf{r}}$, i.e.

$$\sum_{s=1}^{n} (X_s\mathbf{i} + Y_s\mathbf{j} + Z_s\mathbf{k}) = m(\ddot{x}\mathbf{i} + \ddot{y}\mathbf{j} + \ddot{z}\mathbf{k})$$

Therefore

$$\sum X_s = m\ddot{x} \qquad \sum Y_s = m\ddot{y} \qquad \sum Z_s = m\ddot{z} \tag{4.14}$$

Suppose a force $\mathbf{F}$ acts on a particle of mass m for an interval of time t_0 to t_1 during which the velocity of the particle changes from $\mathbf{v}_0$ to $\mathbf{v}_1$. The *impulse* $\mathbf{I}$ of the force $\mathbf{F}$ is defined to be the change of linear momentum produced, i.e.

$$\mathbf{I} = m(\mathbf{v}_1 - \mathbf{v}_0) \tag{4.15}$$

If $\mathbf{F}$ is a variable force,

$$\int_{t_0}^{t_1} \mathbf{F}\, dt = \int_{t_0}^{t_1} m\dot{\mathbf{v}}\, dt = m[\mathbf{v}]_{v0}^{v1} = m(\mathbf{v}_1 - \mathbf{v}_0) = \mathbf{I}$$

so that the impulse of a force is its *time integral* over the interval during which it acts. If $\mathbf{F}$ is a constant force, $\mathbf{I} = \mathbf{F}(t_1 - t_0)$.

A large force acting for a short time is called an *impulsive force* or simply an impulse, and is measured by the change of momentum which it produces.

Let a particle be given a displacement $\delta\mathbf{r}$ in a time δt under the action of a force $\mathbf{F}$. The work done by $\mathbf{F}$ in the time δt is $\mathbf{F}.\delta\mathbf{r}$. Therefore average rate of working of $\mathbf{F}$ in the interval δt is

$$\frac{\mathbf{F}.\delta\mathbf{r}}{\delta t}$$

In the limit as $\delta t \to 0$, the instantaneous rate of working or the *activity* of the force $\mathbf{F}$ is

$$\mathbf{F}.\dot{\mathbf{r}} = \mathbf{F}.\mathbf{v} \tag{4.16}$$

It follows that the work done by the force $\mathbf{F}$ in the interval t_0 to t_1 is given by

$$\int_{t_0}^{t_1} \mathbf{F}.\mathbf{v}\, dt \tag{4.17}$$

The *kinetic energy* T of a particle of mass m moving with a velocity $\mathbf{v}$ is defined by

$$T = \tfrac{1}{2}mv^2 = \tfrac{1}{2}m\mathbf{v}^2 \tag{4.18}$$

$$\frac{dT}{dt} = \tfrac{1}{2}m\frac{d}{dt}(v^2) = m\mathbf{v}.\frac{d\mathbf{v}}{dt} = m\mathbf{a}.\mathbf{v} = \mathbf{F}.\mathbf{v} \tag{4.19}$$

which is the activity of the force acting on the particle. Thus the rate of increase of the kinetic energy of the particle equals the rate of working of the force acting upon it. It follows that the increase of kinetic energy in a finite interval of time is equal to the work done by the force acting upon it during that interval. This statement is true whether the force is constant or variable and is called the *principle of energy*.

4.7 *Motion of a Particle under Gravity*

Let a particle be projected from the point O at time $t = 0$ with a velocity $\mathbf{V}_0 = u_0\mathbf{i} + v_0\mathbf{j}$. Take axes of coordinates Ox, Oy such that $\mathbf{V}_0$ is in the plane

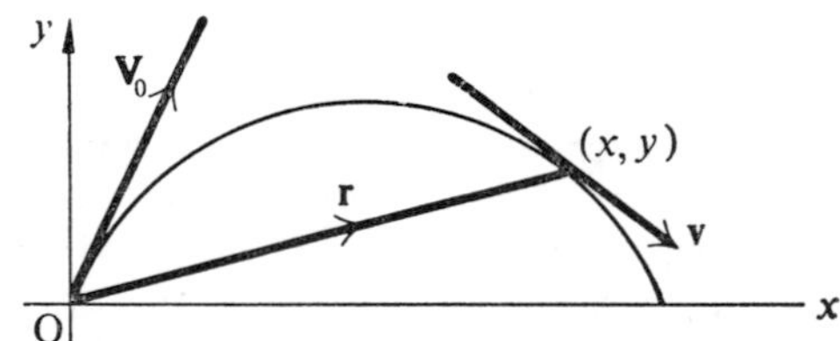

Figure 4.11

xOy and Ox is horizontal. Let $\mathbf{r}$ be the position vector of the particle at time t. Then, if g is acceleration due to gravity vertically downwards,

$$\ddot{\mathbf{r}} = \frac{d\mathbf{v}}{dt} = -g\mathbf{j}$$

Therefore

$$\mathbf{v} = \dot{\mathbf{r}} = -gt\mathbf{j} + \mathbf{V}_0 \text{ satisfying the condition } \mathbf{v} = \mathbf{V}_0 \text{ when } t = 0$$

$$\mathbf{r} = x\mathbf{i} + y\mathbf{j} = -\tfrac{1}{2}gt^2\mathbf{j} + \mathbf{V}_0 t \text{ satisfying the condition } \mathbf{r} = 0 \text{ when } t = 0$$

$$= -\tfrac{1}{2}gt^2\mathbf{j} + t(u_0\mathbf{i} + v_0\mathbf{j})$$

Therefore

$$x = u_0 t \quad \text{and} \quad y = v_0 t - \tfrac{1}{2} g t^2$$

These are the parametric equations of the parabola $y = \dfrac{v_0}{u_0} x - \tfrac{1}{2} g \dfrac{x^2}{u_0^2}$ which has a vertical axis of symmetry.

4.8 *Moment of Momentum (Angular Momentum)*

Let **r** be the position vector of a particle of mass m moving, at a given instant, with a velocity **v**. The velocity vector **v** of the particle, and consequently its momentum vector $m\mathbf{v}$, will be localized vectors coincident with the tangent to the path of the particle at R (Fig. 4.12).

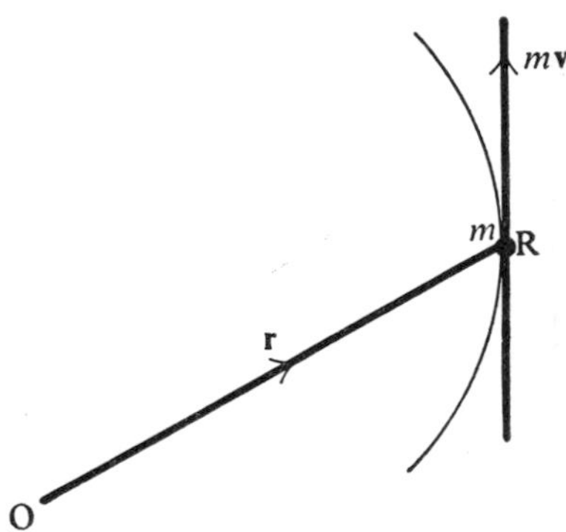

Figure 4.12

The moment of the momentum vector $m\mathbf{v}$ about O is

$$\mathbf{H} = \mathbf{r} \times m\mathbf{v} \tag{4.20}$$

and is also called the *angular momentum* of the particle about O.

Let **F** be the resultant force acting on the particle. The moment or torque of **F** about O is $\mathbf{r} \times \mathbf{F}$. Now

$$\frac{d\mathbf{H}}{dt} = \frac{d}{dt}(\mathbf{r} \times m\mathbf{v}) = \dot{\mathbf{r}} \times m\mathbf{v} + \mathbf{r} \times m\dot{\mathbf{v}} = \mathbf{v} \times m\mathbf{v} + \mathbf{r} \times m\dot{\mathbf{v}}$$

$$= \mathbf{r} \times m\dot{\mathbf{v}} = \mathbf{r} \times \mathbf{F}$$

Therefore the rate of change of the angular momentum of the particle about O equals the moment about O of the resultant force acting on the particle. This is the *Principle of Angular Momentum.*

If the resultant force has a zero moment about O, the angular momentum of the particle about O will remain constant. This is the *Principle of the Conservation of Angular Momentum.*

4.9 Central Forces

Let the position vector of a particle R of mass m with respect to a fixed point O be $\mathbf{r} = \mathbf{OR}$ (Fig. 4.13). The equation of motion of the particle under the

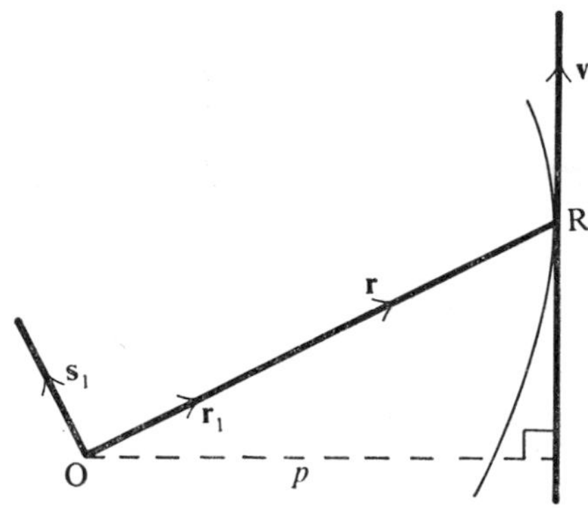

Figure 4.13

action of a force of magnitude $f(r)$ acting in the line OR is

$$m\ddot{\mathbf{r}} = f(r)\mathbf{r}_1 \tag{4.21}$$

where $\mathbf{r}_1$ is a unit vector having the direction of $\mathbf{r}$.

If $f(r) < 0$, then $\ddot{\mathbf{r}}$ has a direction opposite to that of $\mathbf{r}_1$ and the force is directed *towards* O so that the particle is *attracted* towards O. On the other hand, if $f(r) > 0$, the particle is moving under the action of a *repulsive* force from O.

Such forces, always acting in the line OR between a fixed point O and the particle R and having a magnitude which is a function of r only, are called *central forces* having O as the *centre of force.*

Since the moment of a central force about O is zero, it follows from section 4.8 that the angular momentum of the particle about O remains constant. Therefore

$$\mathbf{r} \times m\mathbf{v} = \mathbf{c} \quad \text{a constant vector}$$

$$\mathbf{r} \times \mathbf{v} = \mathbf{h} \quad \text{a constant vector} \tag{4.22}$$

$$= pv\mathbf{k} \tag{4.23}$$

where $\mathbf{k}$ is a unit vector having the direction of $\mathbf{r} \times \mathbf{v}$ and therefore perpendicular to the plane of $\mathbf{r}$ and $\mathbf{v}$, and p is the perpendicular distance of v from O. It follows that the path of the particle lies in a plane determined by the initial vectors $\mathbf{r}$ and $\mathbf{v}$.

By Example 3.3 (p. 71), $\mathbf{v} = \dot{r}\mathbf{r}_1 + r\dot{\theta}\mathbf{s}_1$, therefore by eqn (4.22),

$$\mathbf{r} \times (\dot{r}\mathbf{r}_1 + r\dot{\theta}\mathbf{s}_1) = \mathbf{h} \quad \text{i.e. } r^2\dot{\theta}(\mathbf{r}_1 \times \mathbf{s}_1) = \mathbf{h}$$

$$r^2\dot{\theta}\mathbf{k} = \mathbf{h} = pv\mathbf{k} \tag{4.24}$$

$$r^2\dot{\theta} = pv = h \tag{4.25}$$

Now suppose that the radius vector $\mathbf{r} = \mathbf{OR}$ turns through an angle $\delta\theta$ in a time δt. It sweeps out an area $\frac{1}{2}r^2\,\delta\theta$ in this time. Therefore the rate of description of area by the radius vector is

$$\lim_{\delta t \to 0} \frac{\frac{1}{2}r^2\,\delta\theta}{\delta t} = \tfrac{1}{2}r^2\dot{\theta} = \tfrac{1}{2}h = \tfrac{1}{2}pv \tag{4.26}$$

Thus when a particle moves under a central force, the rate of description of area by the radius vector is constant and equal to $\frac{1}{2}h$.

4.10 Planetary Orbits

Newton's Universal Law of Gravitation states that two bodies of masses M, m are mutually attracted with a force

$$F = \frac{GMm}{r^2}$$

where r is their distance apart and G is a universal constant. Let M be the mass of the sun and m that of a planet, e.g. the earth. Then, if the attractions due to the other planets are assumed to be negligible, the equation of motion of the planet is

$$m\ddot{\mathbf{r}} = -\frac{GMm}{r^2}\mathbf{r}_1 \qquad \text{i.e.} \qquad \ddot{\mathbf{r}} = -\frac{GM}{r^2}\mathbf{r}_1 \tag{4.27}$$

$\mathbf{r}$ being the position vector of the planet with respect to the sun as origin and $\mathbf{r}_1$ a unit vector having the direction of $\mathbf{r}$. By section 4.9, the rate at which the radius vector sweeps out area is constant. This is one of Kepler's Laws of Planetary Motion. By eqns (4.22) and (4.24) we have

$$\mathbf{h} = \mathbf{r} \times \mathbf{v} = r^2\dot{\theta}\mathbf{k}$$

Now

$$\frac{d}{dt}(\mathbf{v} \times \mathbf{h}) = \dot{\mathbf{v}} \times \mathbf{h} = \ddot{\mathbf{r}} \times \mathbf{h} = -\frac{GM}{r^2}\mathbf{r}_1 \times r^2\dot{\theta}\mathbf{k}$$

$$= -GM\dot{\theta}(\mathbf{r}_1 \times \mathbf{k}) = GM\dot{\theta}\mathbf{s}_1 = GM\dot{\theta}\frac{d\mathbf{r}_1}{d\theta} = GM\frac{d\mathbf{r}_1}{dt}$$

Therefore

$$\mathbf{v} \times \mathbf{h} = GM\mathbf{r}_1 + \mathbf{c} \tag{4.28}$$

$$\mathbf{r}.(\mathbf{v} \times \mathbf{h}) = (\mathbf{r} \times \mathbf{v}).\mathbf{h} = GM\mathbf{r}.\mathbf{r}_1 + \mathbf{c}.\mathbf{r}$$

Therefore

$$\mathbf{h}.\mathbf{h} = h^2 = GMr + cr\cos\theta$$

where θ is the angle between the directions of $\mathbf{c}$ and $\mathbf{r}$. Therefore

$$\frac{h^2}{GM} = r\left(1 + \frac{c}{GM}\cos\theta\right) \tag{4.29}$$

Now writing $GM = \mu$, $l = h^2/\mu$ and $e = c/\mu$, the equation of the orbit is

$$l = r(1 + e\cos\theta) \tag{4.30}$$

which is the polar form of the equation of a conic of eccentricity e and semi-latus rectum l referred to its focus as origin. This equation represents an ellipse, parabola, or hyperbola according as e is less than, equal to, or greater than unity. The orbits of the planets are closed curves and are therefore ellipses with the sun at a focus.

Let a, b be the semiaxes of the ellipse; its area is then πab. The rate of description of area by the radius vector drawn from the focus O (Fig. 4.14)

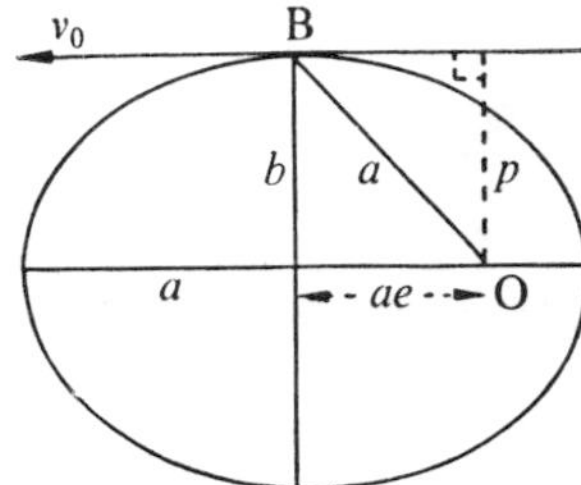

Figure 4.14

is $\frac{1}{2}h$. Therefore the time taken by the planet to complete one circuit of the orbit, the *periodic time* T is given by

$$T = \frac{\pi ab}{\frac{1}{2}h} = \frac{2\pi ab}{\sqrt{(\mu l)}} = \frac{2\pi ab}{\sqrt{\mu}\sqrt{(b^2/a)}} = \frac{2\pi}{\sqrt{\mu}}\, a^{3/2} \tag{14.31}$$

Thus the squares of the periodic times of the planets in their motion relative to the sun are proportional to the cubes of the major axes of their orbits. This law was stated by Kepler following observations of the planets.

From eqn (4.27) we have

$$2\dot{\mathbf{r}}.\ddot{\mathbf{r}} = -\frac{2\mu}{r^3}\,\mathbf{r}.\dot{\mathbf{r}} = -\frac{2\mu}{r^2}\frac{dr}{dt}$$

Integrating with respect to t, we have

$$(\dot{\mathbf{r}})^2 = v^2 = \frac{2\mu}{r} + C \tag{4.32}$$

Let v_0 be the planet's velocity at B, the end of the minor axis, i.e. when $r = a$. Now, if p is the length of the perpendicular from O to the direction of v_0, by eqn (4.25), we have

$$pv = h \quad \text{thus } bv_0 = \sqrt{(\mu l)} = b\sqrt{\left(\frac{\mu}{a}\right)} \quad \text{and} \quad v_0 = \sqrt{(\mu/a)}$$

Thus $v = \sqrt{(\mu/a)}$ when $r = a$. On substituting in eqn (4.32) we have

$$\frac{\mu}{a} = \frac{2\mu}{a} + C \qquad C = \frac{-\mu}{a}$$

Substituting in eqn (4.32), the velocity equation for the orbit is

$$v^2 = \mu\left(\frac{2}{r} - \frac{1}{a}\right) \tag{4.33}$$

4.11 Motion under a Central Force Directly Proportional to Distance

Let a particle of mass m be moving under the action of a force $m\mu\mathbf{r}$ directed towards a fixed point O, $\mathbf{r}$ being the position vector of the particle with respect to O (Fig. 4.15). The equation of motion of the particle is

$$m\ddot{\mathbf{r}} = -m\mu\mathbf{r} \qquad \text{i.e. } \ddot{\mathbf{r}} = -\mu\mathbf{r} \tag{4.34}$$

$$2\ddot{\mathbf{r}}.\dot{\mathbf{r}} = -2\mu\mathbf{r}.\dot{\mathbf{r}}$$

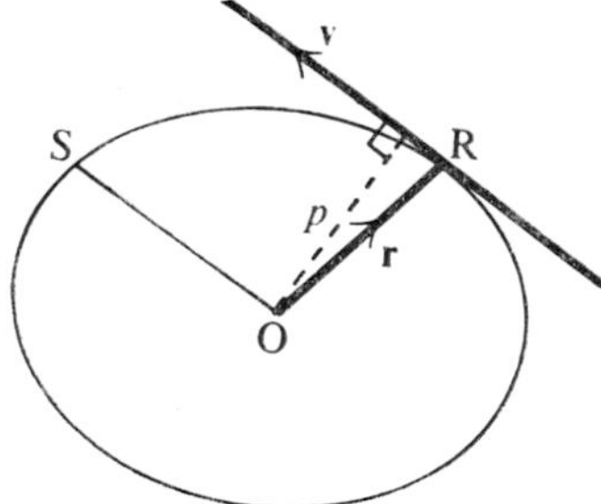

Figure 4.15

Integrating with respect to t,

$$(\dot{\mathbf{r}})^2 = v^2 = -\mu r^2 + C$$

If p is the perpendicular from O to the velocity vector $\mathbf{v}$ at R, we have $pv = h$. Therefore

$$\frac{1}{p^2} = \frac{v^2}{h^2} = \frac{1}{h^2}(C - \mu r^2)$$

Now the p,r equation of an ellipse of semi-axes a,b referred to its centre is

$$\frac{1}{p^2} = \frac{a^2 + b^2 - r^2}{a^2b^2}$$

Hence the path of the particle is an ellipse, centre O such that

$$\frac{\mu}{h^2} = \frac{1}{a^2b^2} \qquad \text{hence } h = ab\sqrt{\mu}$$

and

$$\frac{C}{h^2} = \frac{a^2 + b^2}{a^2b^2} \qquad \text{hence } C = \mu(a^2 + b^2)$$

Therefore

$$v^2 = C - \mu r^2 = \mu(a^2 + b^2 - r^2) \tag{4.35}$$

i.e.

$$v = \sqrt{\mu}\sqrt{(a^2 + b^2 - r^2)} = \text{OS}\sqrt{\mu} \tag{4.36}$$

where OS is the semi-diameter conjugate to OR and parallel to $\mathbf{v}$. The rate of description of area by the position vector $\mathbf{r} = \frac{1}{2}h = \frac{1}{2}ab\sqrt{\mu}$ so that the periodic time T is given by

$$T = \frac{\pi ab}{\frac{1}{2}ab\sqrt{\mu}} = \frac{2\pi}{\sqrt{\mu}} \tag{4.37}$$

and is therefore independent of the size of the orbit.

EXAMPLE 4.11

A particle is projected with a velocity of 120 m/s in a direction inclined at an angle $\tan^{-1} 3/4$ with the horizontal. Find its range on a horizontal plane through the point of projection and the time of flight. (Take $g = 9.8$ m/s^2.)

Solution

$$\ddot{\mathbf{r}} = -g\mathbf{j} \qquad \dot{\mathbf{r}} = -gt\mathbf{j} + \mathbf{a}$$

When $t = 0$,

$$\dot{\mathbf{r}} = 120(\tfrac{4}{5}\mathbf{i} + \tfrac{3}{5}\mathbf{j}) = \mathbf{a} \quad \text{and} \quad \mathbf{a} = 96\mathbf{i} + 72\mathbf{j}$$

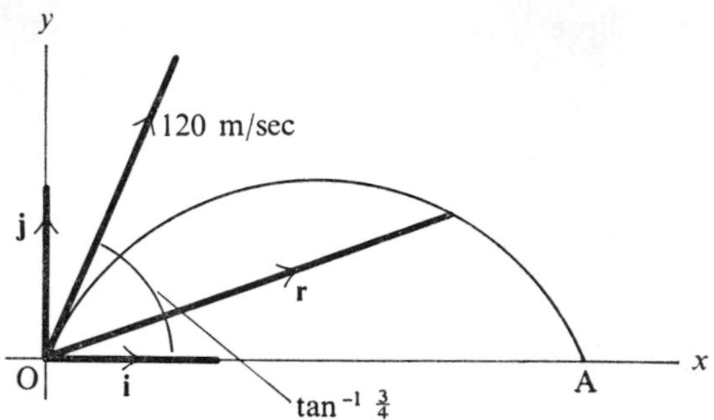

Figure 4.16

Substituting for **a**,

$$\dot{\mathbf{r}} = 96\mathbf{i} + (72 - gt)\mathbf{j}$$

$$\mathbf{r} = 96t\mathbf{i} + (72t - \tfrac{1}{2}gt^2)\mathbf{j}$$

satisfying the initial condition $\mathbf{r} = 0$ when $t = 0$. Now $\mathbf{r} = \mathbf{OA}$ when $72t - \frac{1}{2}gt^2 = 0$, i.e. when $t = 144/g = 14.69$ sec, so that the time of flight is $144/g$ sec and the range is

$$OA = |\mathbf{r}|_{t=144/g} = 96 \times 144/g = 1\,411 \text{ m approx}$$

EXAMPLE 4.12
A particle moves under the attraction of a force μ/r^2 per unit mass directed towards a fixed point S. The distance of the particle from S is r (Fig. 4.17). The particle is projected from a point P, where SP $= a$, with a velocity $\sqrt{(\mu/a)}$ in a direction which makes an angle $\frac{1}{4}\pi$ with SP. Show that it moves in an elliptic orbit of eccentricity $1/\sqrt{2}$ in a periodic time $(2\pi a^{3/2})/\mu^{1/2}$.

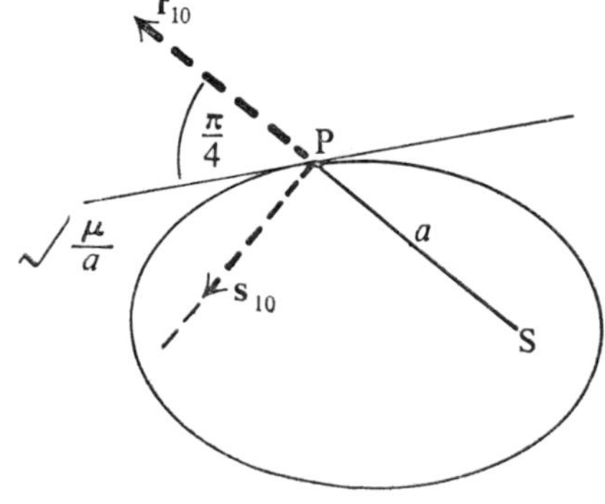

Figure 4.17

Solution

$$\mathbf{h} = r^2\dot{\theta}\mathbf{k} = \sqrt{\left(\frac{\mu}{a}\right)}\frac{1}{\sqrt{2}}\,a\mathbf{k} = \sqrt{(\tfrac{1}{2}\mu a)}\mathbf{k}$$

Now $\ddot{\mathbf{r}} = -(\mu/r^2)\,\mathbf{r}_1$. Therefore by eqn (4.28),

$$\mathbf{v} \times \mathbf{h} = \mu\mathbf{r}_1 + \mathbf{c}$$

Initially $\mathbf{v} = \sqrt{(\mu/a)}(\cos\tfrac{1}{4}\pi\mathbf{r}_{10} + \sin\tfrac{1}{4}\pi\mathbf{s}_{10}) = \sqrt{(\mu/2a)}(\mathbf{r}_{10} + \mathbf{s}_{10})$ where $\mathbf{r}_{10}$ and $\mathbf{s}_{10}$ are the radial and transverse unit vectors at P. Therefore, initially,

$$\mathbf{v} \times \mathbf{h} = \tfrac{1}{2}\mu(\mathbf{r}_{10} + \mathbf{s}_{10}) \times \mathbf{k} = \tfrac{1}{2}\mu(-\mathbf{s}_{10} + \mathbf{r}_{10}) = \mu\mathbf{r}_{10} + \mathbf{c}$$

$$\mathbf{c} = -\tfrac{1}{2}\mu(\mathbf{r}_{10} + \mathbf{s}_{10}) \qquad c = |\mathbf{c}| = \tfrac{1}{2}\mu\sqrt{2} = \mu/\sqrt{2}$$

Hence, the eccentricity of the orbit is

$$e = c/\mu = 1/\sqrt{2}$$

Thus the orbit is an ellipse of eccentricity $1/\sqrt{2}$.
Now

$$l = \frac{h^2}{\mu} = \frac{1}{\mu}\frac{\mu a}{2} = \frac{a}{2}$$

Let A and B be the semi-major and semi-minor axes respectively, then

$$a/2 = l = B^2/A = A(1 - e^2) = A/2 \qquad \text{thus } A = a$$

By eqn (4.31), the periodic time is $2\pi A^{3/2}/\mu^{1/2} = (2\pi a^{3/2})/\mu^{1/2}$.

EXAMPLE 4.13
A particle of unit mass is describing an elliptic orbit with semi-axes $2a$ and a under an attraction μr to the centre C when it is at a distance r from C. If the intensity of the attraction is suddenly increased in the ratio 4:1 when the particle is at a distance $3a/2$ from C, show that the semi-axes of the new orbit are $\tfrac{1}{8}a(\sqrt{79} \pm \sqrt{15})$.

Solution By eqn (4.35)

$$v^2 = \mu(a^2 + b^2 - r^2) = \mu(5a^2 - r^2)$$

Let v_1 be the velocity when $r = 3a/2$ (Fig. 4.18). Therefore

$$v_1^2 = \mu(5a^2 - 9a^2/4) = 11\mu a^2/4$$

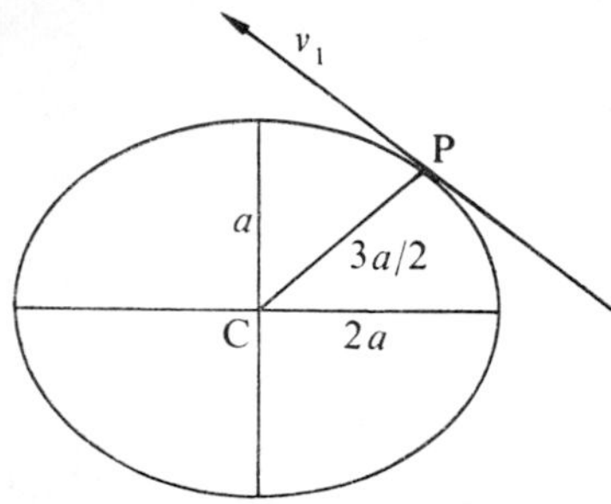

Figure 4.18

When the particle is at P, the attraction is suddenly quadrupled so that "μ" becomes 4μ and the velocity equation for the new elliptic orbit is

$$v^2 = 4\mu(A^2 + B^2 - r^2)$$

where A, B are the semi-axes of the new orbit.

When $r = 3a/2$, we have $v = v_1$, therefore

$$11\mu a^2/4 = 4\mu(A^2 + B^2 - 9a^2/4) \qquad A^2 + B^2 = 47a^2/16$$

When the force is suddenly quadrupled at P, the moment of the velocity about C is not changed. Therefore

$$h = 2a \times a\sqrt{\mu} = 2(\sqrt{\mu})AB \qquad AB = a^2$$

Therefore

$$(A \pm B)^2 = \tfrac{47}{16}a^2 \pm 2a^2 = \tfrac{79}{16}a^2 \text{ and } \tfrac{15}{16}a^2$$

i.e. $A + B = \frac{1}{4}a\sqrt{79}$ and $A - B = \frac{1}{4}a\sqrt{15}$

$$A = \tfrac{1}{8}a(\sqrt{79} + \sqrt{15}) \qquad B = \tfrac{1}{8}a(\sqrt{79} - \sqrt{15})$$

EXERCISES 4.5

1. The acceleration of a particle P moving on the curve $r = ae^{k\theta}$ always has the direction of the line OP where O is the origin. Show that the magnitude of the acceleration is proportional to r^{-3}.
2. A particle is projected with a velocity of 130 m/s at an angle $\tan^{-1} 5/12$ with the horizontal. Show that its range on a horizontal plane through the point of projection is $12\,000/g$ metres and the time of flight is $100/g$ sec.

If the particle is projected directly up a plane which passes through the point of projection and is inclined at an angle $\tan^{-1} 1/4$ above the horizontal, show that it strikes the plane after $40/g$ sec, with a velocity $10\sqrt{145}$ m/s whose direction is inclined at $\tan^{-1} 1/12$ to the horizontal. Show also that the range on the plane is $1\,200\sqrt{17}/g$ metres.

3. Two particles are projected simultaneously with velocities u and v from points A and B respectively, each directly towards the other. Show that the particles will collide at a point vertically below a point C of AB such that AC/CB $= u/v$.

4. A particle is attracted towards a fixed point S by a force $3ku^2/r^2$ per unit mass (k, u constants) where r is its distance from S. If the particle is projected from a point A, distant k from S, with a velocity u whose direction makes an angle $\pi/3$ with **SA**, show that it will describe an ellipse of eccentricity $\sqrt{(7/12)}$.

5. A particle is describing an elliptic orbit about a centre of force at a focus S. At a point P in its orbit, distant c from S, the tangent to the orbit is inclined at an angle $\sin^{-1} 3/5$ to **SP**. At a point Q on the orbit, distant $3c/2$ from S, the speed of the particle is half its speed at P. If the intensity of the attraction per unit mass at unit distance is μ, find the speed at P. Show that the major axis of the orbit is $9c/5$ and find its eccentricity. [$\frac{2}{3}\sqrt{(2\mu/c)}$, $\sqrt{(\frac{29}{45})}$]

6. A particle moves under a central force inversely proportional to the square of the distance from a fixed point S, the accelerating effect of the force at a distance of 1 m being $g = 9.8$ m/s^2. If the particle is projected in a direction perpendicular to a radius at a distance 3/4 m from S with a velocity of 4 m/s, determine the major axis and the eccentricity of the orbit. What value must the velocity of projection exceed to make the orbit a hyperbola? [0.97 m, 0.224, 5.11 m/s]

7. A point P moves so that its velocity is the vector sum of two components of constant magnitudes u, v, the first being in a fixed direction and the second perpendicular to the line joining P to a fixed point S. Show that the orbit is a conic of eccentricity u/v.

8. A particle of mass m is describing an ellipse of major and minor axes $2a$, $2b$ respectively about a centre of force at the centre. When it reaches the end of the major axis, it strikes and coalesces with a particle of mass nm at rest. The central attraction per unit mass is unchanged. Prove that the new orbit is an ellipse of major and minor axes $2a$, $2b/(n + 1)$ respectively.

5 *Probability theory*

5.1 *Introduction*

The concept of probability, which has played a fundamental part in the development of the theory of statistics, arose from the consideration of games of chance which are essentially experiments of a repetitive nature. In any game of chance, such as tossing a coin, throwing a die or drawing a card, the outcome of any particular trial is uncertain. Nevertheless, experience has shown that games of chance and many repetitive industrial operations and scientific experiments behave, in the long run, as if they were essentially stable. For example, an unbiased coin would show heads in about one-half of a large number of tosses. Again, whilst an insurance company could not predict which particular man would die at the age of 60, it could predict what proportion of men would die at that age.

To enable the statistician to predict the outcomes of future trials of a repetitive experiment and to study its properties, it is essential that he should construct a mathematical model applicable to the experiment.

5.2 *Outcomes and Events*

Any repetitive experiment has a number of alternative *outcomes* which are *mutually exclusive*. For example, when a coin is tossed, there are two alternative outcomes—head or tail (H or T). If the coin is unbiased, H or T may be expected to fall with approximately equal frequencies in a large number of trials. Consequently, the probability of each of the outcomes H or T is said to be 1/2.

If two coins are tossed (or one coin is tossed twice), there are four possible outcomes which are mutually exclusive, namely

$HH \quad HT \quad TH \quad TT$

which are equally likely. Consequently, for balanced coins, each of these outcomes is given the probability 1/4.

When an unbiased die is thrown, there are six possible mutually exclusive outcomes, 1, 2, 3, 4, 5, or 6, the probability of each outcome being 1/6. If two true dice are thrown, there will be 6×6, i.e. 36 outcomes:

(1, 1)	(1, 2)	(1, 3)	(1, 4)	(1, 5)	(1, 6)
(2, 1)	(2, 2)	(2, 3)	(2, 4)	(2, 5)	(2, 6)
(3, 1)	(3, 2)	(3, 3)	(3, 4)	(3, 5)	(3, 6)
(4, 1)	(4, 2)	(4, 3)	(4, 4)	(4, 5)	(4, 6)
(5, 1)	(5, 2)	(5, 3)	(5, 4)	(5, 5)	(5, 6)
(6, 1)	(6, 2)	(6, 3)	(6, 4)	(6, 5)	(6, 6)

each outcome occurring with a probability of 1/36.

A single outcome is called a *simple event* or an *element* but an *event* may comprise either one outcome or a group of outcomes, for example

(*a*) When two coins are tossed, the event in which one head and one tail occur together comprises the two outcomes HT and TH, whilst the event in which two tails occur consists of the single element TT.

(*b*) When two dice are thrown, the event in which two fours occur consists of the single outcome (4, 4) whilst the event, in which a total of eight is thrown, comprises the five elements

(6, 2) (5, 3) (4, 4) (3, 5) (2, 6)

From such considerations, we derive the following classical definition of probability:

DEFINITION 5.1

If a repetitive experiment has n mutually exclusive and equally likely outcomes*, and if n_A of the outcomes comprise an event A, then the probability of the event A is n_A/n. $(0 \leqslant n_A/n \leqslant 1)$

* Probability distributions for which all the outcomes are equally likely are called "Laplace distributions".

With reference to (*a*) and (*b*), it follows that:

in (*a*) the probability of an event comprising one head and one tail is $2/4 = 1/2$

and of the event comprising two tails $1/4$

and in (*b*) the probability of throwing two fours is $1/36$ and of throwing a total of eight $5/36$.

In applying the above definition, it is important to bear in mind the requirement that outcomes must be *equally likely* and *mutually exclusive*. Note, for example, that the probability of drawing either a king or a diamond from a pack of cards is not 17/52 but 16/52 since one of the four kings is also one of the thirteen diamonds.

Probabilities derived from Definition 5.1 are called "a priori" probabilities being obtained by deduction from idealized models, e.g. from a "fair" die.

As already observed, the outcomes of many repetitive experiments show stability in the long run. Suppose that a series of experiments be performed under conditions which are as similar as possible and that the frequency of the occurrence of an event A be recorded.

Let us now postulate that the relative frequency of the event A is an approximation to the probability of the event A. A probability measure determined in this manner is known as an "a posteriori" probability or a statistical probability.

5.3 Sample Points and Sample Space

Suppose that every possible outcome of an experiment can be enumerated and a probability assigned to each outcome. It is convenient to represent each outcome by a point specified in appropriate coordinates—such points are called *sample points* to each of which a probability may be assigned, as in the following examples.

The two outcomes H and T which result from the tossing of a true coin may conveniently be represented by two sample points on a straight line, 0 for the tail and 1 for the head.

Similarly, the six outcomes arising from throws of a symmetrical die may be represented by six sample points 1, 2, 3, . . . , 6 on a line each having a probability 1/6.

The four outcomes HH, HT, TH, TT resulting from throws of two unbiased coins may be conveniently represented by the four sample points (1, 1), (1, 0), (0, 1), (0, 0) in two-dimensional cartesian space, each having a probability 1/4 (Fig. 5.1).

If three balanced coins are tossed, the eight outcomes could be represented by eight sample points in three dimensional space, each point having a probability 1/8.

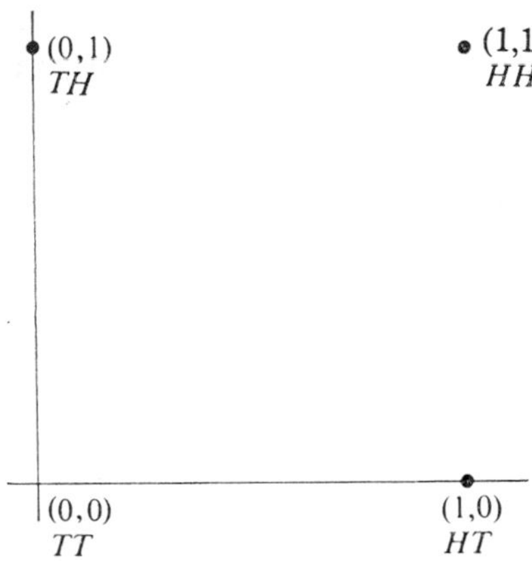

Figure 5.1

When two fair dice are thrown, the 36 outcomes (Section 5.2) may be represented by the 36 sample points illustrated in Fig. 5.2, each bearing a probability 1/36.

In the above examples, the outcomes are "equally likely" so that equal probabilities would be assigned to the associated sample points. In general, however, the probabilities assigned to sample points on the basis of the relative frequencies of the various outcomes in a large series of trials would be unequal.

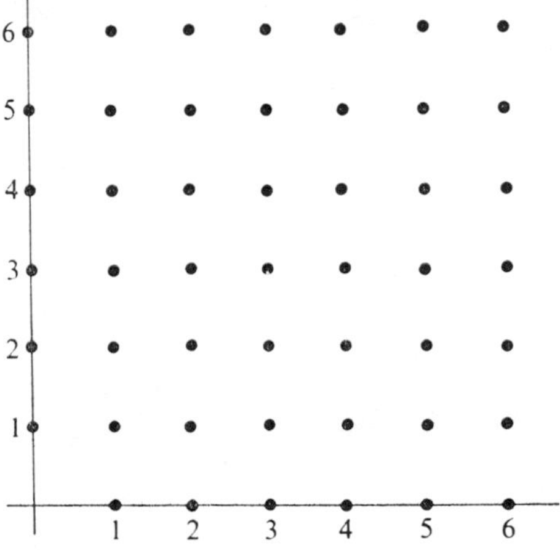

Figure 5.2

DEFINITION 5.2
The set of sample points representing all possible outcomes of an experiment comprises the *sample space* of that experiment.

A sample space is said to be discrete or continuous according to whether it contains a finite number or a continuum of sample points.

The probability assigned to any sample point is necessarily positive and the sum of the probabilities of all the sample points comprising the sample space of an experiment is 1.

Let the probabilities assigned to the n sample points representing the n outcomes of an experiment be $p_1, p_2, \ldots, p_n$ then

$$0 \leqslant p_i \leqslant 1 \quad \text{and} \quad \sum_{i=1}^{n} p_i = 1 \qquad (i = 1, 2, 3, \ldots, n)$$

5.4 *The Probability of an Event*

Suppose that for each outcome of a repetitive experiment, it can be decided unambiguously whether or not an event A has occurred so that each sample point is one for which the event A has or has not occurred. The probability of the event A is defined as follows:

DEFINITION 5.3
The sum of the probabilities assigned to all those sample points associated with the occurrence of the event A is the probability $p(A)$ of the event A. Let the Venn diagram (Fig. 5.3) represent the sample space for such an experiment and let the set A comprise those sample points associated with the occurrence of the event A. The sum of the probabilities assigned to these points gives the probability $p(A)$ of the event A, that is

$$p(A) = \sum_{A} p_i$$

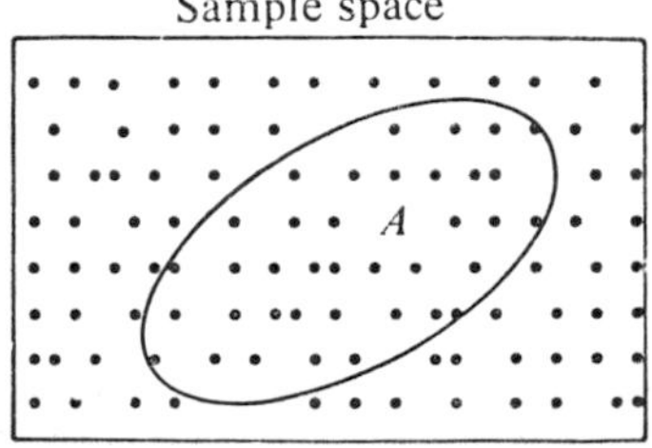

Figure 5.3 The set A of sample points associated with the event A.

where the sum is taken over the sample points in set A. It is clear that

$$0 \leqslant p(A) \leqslant 1$$

and the probability that the event A does not happen is

$$p(\bar{A}) = 1 - p(A)$$

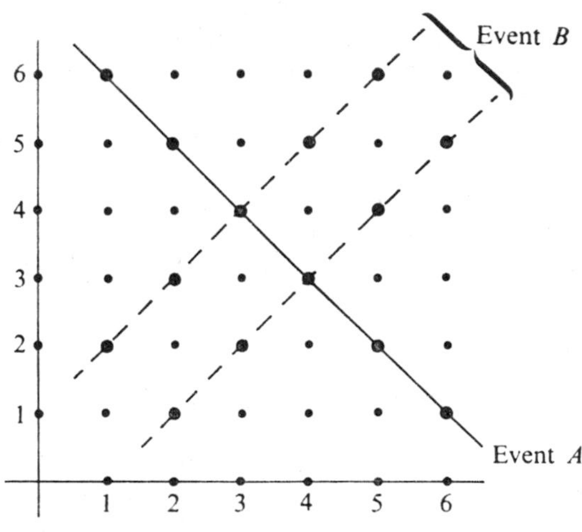

Figure 5.4

Figure 5.4 shows the 36 sample points which comprise the sample space for throws of two symmetrical dice. The probability of

(i) Event A, the throwing of a total of 7, is $6/36 = 1/6$.

(ii) Event B, the throwing of a difference of 1, is $10/36 = 5/18$.

The outcomes (4, 3) and (3, 4) are common to the events A and B. Consequently, the probability that either the event A or the event B or both occur is $14/36 = 7/18$, whilst the probability that both occur, i.e. the probability of a throw for which the sum is 7 and the difference is 1, is $2/36 = 1/18$.

5.5 *Combinatorial Formulae*

To facilitate the computation of the probability of an event by counting the sample points associated with the occurrence of that event in accordance with Definition 5.1, the combinatorial formulae of Algebra are applied. The development of these formulae is based upon the following axioms. (1) If two mutually exclusive events A and B can occur in m ways and in

n ways respectively then

(i) either the event A or the event B can occur in $(m + n)$ ways
(ii) the events A and B can occur simultaneously in mn ways.

For example, let the mutually exclusive events be

A the drawing of an ace from a pack of cards—this can be done in 4 ways
B the drawing of a king from a pack of cards—this can also be done in 4 ways.

(i) If one card only is drawn, the number of ways in which either an ace or a king may be drawn is $4 + 4$, i.e. 8 ways.
(ii) If two cards are drawn, the number of ways in which one ace and one king may be drawn is 4×4, i.e. 16 ways, since any one of the four aces may be drawn with any one of the 4 kings.

(2) For three mutually exclusive events A, B and C which can occur in m, n and p ways respectively

(i) either A or B or C can occur in $(m + n + p)$ ways
(ii) A, B and C can occur simultaneously in mnp ways.

These axioms may be generalized, in an obvious manner, to apply to more than three events.

Let us now determine the number of ways in which a group of n different objects may be arranged. Any such arrangement of the n objects *in a definite order* is called a *permutation.*

Any one of the n objects may be placed in the first position which may therefore be filled in n ways. The filling of the first position is thus an event which may occur in n ways. Having placed any one object in the first position, $(n - 1)$ objects remain so that the filling of the second place—the second event—may occur in $(n - 1)$ ways. The first and second positions may therefore be filled in $n(n - 1)$ ways. Having placed any two objects in the first two positions, $(n - 2)$ objects remain and the third position may be filled in $(n - 2)$ ways and so on.

It follows that the number of permutations of n objects amongst themselves is

$$n(n - 1)(n - 2) \cdots 2 \times 1 = n! \tag{5.1}$$

where $n!$ is called factorial n.

Let us now determine the number of permutations of n objects when only r $(<n)$ of the objects are used in any permutation. As before, the various positions may be filled successively in n, $(n - 1)$, $(n - 2)$, . . . ways but when we reach the final, i.e. the rth position in the permutation, $(r - 1)$ objects have been placed so that $n - (r - 1)$ objects remain. Consequently, the rth position may be filled in $\{n - (r - 1)\}$ ways.

Therefore the number of permutations of n different objects taken r at a time, which will be denoted by the symbol ${}^nP_r$, is given by

$$ {}^nP_r = n(n-1)(n-2)\cdots(n-r+1) = \frac{n!}{(n-r)!} \tag{5.2} $$

The number of *combinations* of n different objects taken r at a time is defined to be the number of different selections, each comprising r objects, which can be chosen from n objects without reference to their order within a selection. This quantity will be denoted by $\binom{n}{r}$. (The symbol ${}^nC_r$ is also used.)

Two combinations are different when they do not contain the same objects, e.g. *abc* and *abd* are different three-letter combinations but *abc*, *acb*, *bca*, *bac*, *cab*, *cba*, are the 6 (i.e. 3!) permutations of the same combination.

Each combination of r objects may be permuted in $r!$ ways. Thus

$$ \binom{n}{r} x r! = {}^nP_r $$

$$ \binom{n}{r} = \frac{{}^nP_r}{r!} = \frac{n(n-1)(n-2)\cdots(n-r+1)}{r!} = \frac{n!}{r!\,(n-r)!} \tag{5.3} $$

It easily follows from (5.3) that

$$ \binom{n}{r} = \binom{n}{n-r} \tag{5.4} $$

This result is otherwise obvious since eqn (5.3) may be interpreted as the number of ways in which n different objects may be divided into two groups containing respectively r and $(n-r)$ objects.

Let us now determine the number of ways in which n objects may be divided into three groups containing respectively n_1, n_2 and n_3 objects such that $n_1 + n_2 + n_3 = n$. First, divide up the n objects in two groups containing n_1 and $(n_2 + n_3)$ objects respectively.

By (5.4), this may be done in $\binom{n}{n_1}$ ways.

Now divide any group containing $(n_2 + n_3)$ objects into two groups containing respectively n_2 and n_3 objects.

This may be done in $\binom{n_2 + n_3}{n_2}$ ways.

It follows that the number of ways in which the complete process of subdivision, resulting in three groups containing respectively n_1, n_2 and n_3

objects, may be performed is

$$\binom{n}{n_1}\binom{n_2+n_3}{n_2} = \frac{n!}{n_1!\,(n_2+n_3)!}\,\frac{(n_2+n_3)!}{n_2!\,n_3!} = \frac{n!}{n_1!\,n_2!\,n_3!} \tag{5.5}$$

By extending the above argument, it easily follows that the number of ways in which n objects may be divided into k groups containing respectively $n_1, n_2, n_3, \ldots, n_k$ objects such that $n_1 + n_2 + n_3 + \cdots + n_k = n$ is

$$\frac{n!}{n_1!\,n_2!\,n_3!\cdots n_k!} \tag{5.6}$$

Equation (5.6) may be also interpreted as the number of permutations of n objects containing k groups comprising n_1 identical objects of one type, n_2 identical objects of a second type and so on to n_k identical objects of a kth type.

The total number of different permutations, say P, in this case would be less than $n!$, the number of permutations if all the objects were different. Each of these P permutations would give rise to additional permutations if objects of the same type were made different, If, for example, the n_1 objects of the first type were made different, they could be rearranged in $n_1!$ ways within each of the P permutations giving rise to $n_1!P$ permutations. Similarly if the n_2 objects were all made different, $n_2!$ times as many permutations as before, i.e. $n_1!\,n_2!\,P$ permutations, would result. Continuing this procedure, until all the objects have been made different, the total number of permutations would be

$$n_1!\,n_2!\ldots n_k!\,P = n! \qquad \text{i.e. } P = \frac{n!}{n_1!\,n_2!\ldots n_k!}$$

EXAMPLE 5.1

A set of m parallel lines intersect n parallel lines, having a different direction, to form a network of parallelograms. How many parallelograms are formed?

Solution On any one line of the second set, there are m points of intersection by lines of the first set. The number of pairs of such points is $\binom{m}{2}$. Similarly, on any line of the first set, there are $\binom{n}{2}$ pairs of points of intersection by the second set. Therefore

The number of parallelograms in the grid $= \binom{m}{2}\binom{n}{2} = mn(m-1) \times (n-1)/4$.

EXAMPLE 5.2

If p_1 and p_2 are the probabilities of two independent events, show that the probability of the simultaneous occurrence of these two events is p_1p_2.

In 18 games of chess, A wins 8, B wins 6, and 4 are drawn. A and B play a tournament of 3 games. On the basis of these data, estimate the probability that

(*a*) A wins all three games (*b*) A and B win alternately
(*c*) Two games are drawn (*d*) A wins at least one game.

What are the odds against A losing the first two games to B? (*A.E.B. A-level*)

Solution

The probability that A will win, $p(A) = 8/18 = 4/9$
The probability that B will win, $p(B) = 6/18 = 1/3$
The probability that a game will be drawn, $p(C) = 4/18 = 2/9$

(*a*) The probability that A wins the first, second and third games is

$$p(A_1A_2A_3) = p(A_1)p(A_2)p(A_3) = (4/9)^3 = 64/729$$

(*b*) The probability of the mutually exclusive events $A_1B_2A_3$ and $B_1A_2B_3$ is

$$p(A_1B_2A_3) + p(B_1A_2B_3) = (\tfrac{4}{9} \times \tfrac{1}{3} \times \tfrac{4}{9}) + (\tfrac{1}{3} \times \tfrac{4}{9} \times \tfrac{1}{3}) = 28/243$$

(*c*) The probability that two games are drawn and one game is not drawn is

$$3 \times p(D_1\, D_2\, \bar{D}_3) = 3 \times (\tfrac{2}{9})^2 \times \tfrac{7}{9} = 28/243$$

(*d*) The probability that A wins *at least* one game is

$$1 - p(A \text{ does not win a game}) = 1 - p \text{ (draws and wins by } B \text{ only)}$$

$$= 1 - (\tfrac{5}{9})^3 = 604/729$$

The probability $p(\bar{A}_1\bar{A}_2)$ that A loses the first two games to B includes the probability that A loses the first two games and wins the third game *plus* the probability that A loses the first two games and draws the third *plus* the probability that A loses all three games.

$$p(\bar{A}_1\bar{A}_2) = p(B_1B_2A_3) + p(B_1B_2D_3) + p(B_1B_2B_3)$$

$$= (\tfrac{1}{3})^2\tfrac{4}{9} + (\tfrac{1}{3})^2\tfrac{2}{9} + (\tfrac{1}{3})^3 = \tfrac{1}{9}$$

Thus, the odds *against* A losing the first two games is 8:1.

EXAMPLE 5.3

5-card hands are drawn from a pack. What is the probability of drawing (*a*) precisely two aces, (*b*) at least two aces?

Solution In drawing 5-card hands, the possible number of outcomes is $\binom{52}{5}$

(*a*) Number of aces in the pack = 4

Number of ways in which a pair of aces may be drawn = $\binom{4}{2}$

The other three cards in a hand occur as selections of 3 drawn from the other 48 cards and therefore may be drawn in $\binom{48}{3}$ ways.

Therefore 5-card hands containing 2 aces occur in $\binom{4}{2}\binom{48}{3}$ ways.
Therefore, required probability

$$= \frac{\binom{4}{2}\binom{48}{3}}{\binom{52}{5}}$$

$$= \frac{\dfrac{4!}{2!\,2!} \times \dfrac{48!}{3!\,45!}}{\dfrac{52!}{5!\,47!}} = \frac{2\,162}{54\,145}$$

(*b*) The hands must contain 2, 3 or 4 aces. The alternatives to this are that the hands contain either 0 or 1 ace.

Hands containing no aces occur in $\binom{48}{5}$ ways

Hands containing 1 ace occur in $\binom{4}{1}\binom{48}{4}$ ways

Therefore probability of the alternatives

$$= \frac{\binom{48}{5} + \binom{4}{1}\binom{48}{4}}{\binom{52}{5}} = 51\,888/54\,145$$

Probability of at least two aces = 1 − 51 888/54 145 = 2 257/54 145

EXAMPLE 5.4

During one term, a college team plays 6 cricket matches which result in 3 wins, 2 losses and 1 draw. In how many different ways is this possible?

Solution The required answer is the number of ways in which the 6 opposing teams may be partitioned into 3 groups containing respectively 3 losing teams, 2 winning teams and 1 drawing team. Therefore

$$\text{Required number of ways} = 6!/(3!\,2!\,1!) = 60$$

EXAMPLE 5.5

6 balls are thrown into 4 boxes so that each ball falls into one of the boxes and is equally likely to fall in any one of the boxes. Find the probability that the fourth box contains precisely two balls.

Solution Each ball may fall, with equal probability, in any one of the four boxes.

For 6 balls, the total number of outcomes is 4^6.

The number of ways in which two balls may be selected for the fourth box is $\binom{6}{2}$.

For any such selection, the remaining 4 balls may be distributed amongst the remaining 3 boxes in 3^4 ways.

Therefore

The number of favourable outcomes $= \binom{6}{2} 3^4$

Required probability $= \binom{6}{2} 3^4/4^6 = 0.2966$ approx.

EXERCISES 5.1

1. How many different groups of results are possible for 10 football matches? [59 049]
2. Show that n persons may be seated around a circular table in $(n-1)!$ ways.
3. How many integers are factors of the number $2^6 \times 3^4 \times 5^2 \times 7 \times 11$ not counting 1 or the number itself? [418]
4. In how many ways can a pair of triangles be drawn with 6 given points as vertices, no three of the points being collinear? [10]
5. In how many ways may 5 red counters, 4 white and 2 black be arranged in a row? [6 930]
6. Two dice are thrown. What are the probabilities that the total score will be (*a*) 5, (*b*) 0, (*c*) 10, (*d*) 14, (*e*) less than 13? What is the probability of a score of either 6 or a double? [(*a*) 1/9, (*b*) 0, (*c*) 1/12, (*d*) 0, (*e*) 1; 5/18]
7. What is the probability of throwing not more than 4 with a die? When two dice are thrown, what is the probability of scoring either 7 or 11? [2/3, 2/9]
8. 4 cards are drawn at random from a pack of cards. What is the probability that they are honours? (A, K, Q, J, 10 are the honours cards.) [57/3 185]

9. What is the probability that 5 cards drawn from a pack are all of the same suit? [33/16 660]
10. A hand of 5 cards is dealt from a well-shuffled pack. What is the probability that the hand consists of 5 cards in sequence but not necessarily of the same suit? [Hint: There are 9 sequences in 13 different cards.] [192/54 145]
11. A bag contains 9 balls of which 2 are red, 3 white and 4 black. 3 balls are drawn at random from the bag. What is the probability that
    (*a*) the 3 balls are of different colours?
    (*b*) 2 balls are of the same colour and the third of a different colour?
    (*c*) the 3 balls are of the same colour? [(*a*) 2/7, (*b*) 55/84, (*c*) 5/84]
12. A bag contains 3 white, 4 red and 6 black balls. Two balls are drawn. How many possible outcomes are there? In how many of these will the two balls be of the same colour. What is the probability that the two balls will be of different colours? [78, 24, 9/13]
13. Find the number of permutations of the following symbols
    (*a*) A, B, C, D, E (*b*) A, A, C, D, E (*c*) A, A, B, B, B
    [(*a*) 120, (*b*) 60, (*c*) 10]
14. 15 students register to take a course which is provided at 3 different times.
    (*a*) In how many different ways could the students be assigned to three classes.
    (*b*) In how many different ways can the students be divided equally between 3 classes? [(*a*) 3^{15}, (*b*) 756 756]
15. 9 students are to be assigned to 3 rooms, each capable of accommodating 3 students. In how many ways may this be done? If 2 particular students refuse to share a room, in how many ways may the students be accommodated? [1 680, 1 260]
16. 10 applicants for an appointment are interviewed by 3 persons who independently place the candidates in order of merit. It is decided to appoint that candidate who is placed first by at least 2 of the 3 interviewers. Estimate the probability of the appointment of some particular candidate. [0.28]

5.6 *Expectation or Expected Value*

Suppose that a variable x assumes n values x_i with the respective probabilities p_i $(i = 1, 2, \ldots, n)$ in n mutually exclusive outcomes of a trial. The n pairs of values x_i, p_i constitute the probability distribution of the discrete random variable x.

If x is a *continuous* variable, one cannot speak of the probability of a particular value of x. Instead we consider the probability that the value of the variable lies in an infinitesimal interval $x - \frac{1}{2}dx$ to $x + \frac{1}{2}dx$ and restrict our attention to those cases in which this probability is expressible in the form $p(x)dx$ where $p(x)$ is a continuous function. $p(x)$ is called the *probability density function* (p.d.f.) or probability function.

DEFINITION 5.4
The *expectation* or *expected value* of the *discrete* random variable x is defined to be

$$E(x) = \sum_{i=1}^{n} p_i x_i \tag{5.7}$$

If x is a *continuous* variable having a *continuous* probability density function $p(x)$, the expectation of x is defined to be*

$$E(x) = \int_{-\infty}^{+\infty} xp(x)\,dx \tag{5.8}$$

For any such probability distribution, the expected value $E(x)$ may be interpreted as the average or *mean value* of x and will be denoted by μ.

EXAMPLE 5.6
(A) In a gambling game, a player rolls a die and is paid as many pounds as the number he throws. What entrance fee should he pay for a fair game?

Solution The player would expect to win

$$\pounds\{\tfrac{1}{6}(1) + \tfrac{1}{6}(2) + \cdots + \tfrac{1}{6}(6)\} = \pounds 3.50$$

so that he would expect to cover an entrance fee of £3.50, in the long run, by his winnings.

(B) An urn contains 3 red balls and 10 black. 4 balls are drawn together. Find the probability distribution of x, the number of red balls drawn, and the expected value of x.

Solution When 4 black balls are drawn, $x = 0$. Therefore
Probability that $x = 0$ is

$$\binom{10}{4} \Big/ \binom{13}{4} = 42/143$$

Probability of drawing 3 black balls and 1 red (i.e. of $x = 1$) is

$$\binom{10}{3}\binom{3}{1} \Big/ \binom{13}{4} = 72/143$$

* If $p(x)$ is defined for a limited range of x, eqn (5.8) is still valid since $p(x)$ is taken to be zero outside this range.

The probability of drawing, 2 black balls and 2 red (i.e. of $x = 2$) is

$$\binom{10}{2}\binom{3}{2}\bigg/\binom{13}{4} = 27/143$$

The probability of drawing, 1 black ball and 3 red (i.e. of $x = 3$) is

$$\binom{10}{1}\binom{3}{3}\bigg/\binom{13}{4} = 2/143$$

Note that, as expected, the sum of these probabilities is 1. The mean number of red balls drawn is

$$\begin{aligned} E(x) &= 0 \times 42/143 + 1 \times 72/143 + 2 \times 27/143 + 3 \times 2/143 \\ &= 132/143 \end{aligned}$$

(C) In an indefinite series of independent trials, there is a constant probability p of success. Show that the expectation of the number of failures preceding the first success is $p^{-1} - 1$.

Solution Let $q = (1 - p)$ be the constant probability of a failure and let x be the number of failures preceding the first success. Therefore

qp = the probability of 1 failure followed by the first success
q^2p = the probability of 2 failures followed by the first success
q^3p = the probability of 3 failures followed by the first success and so on.

$$\begin{aligned} E(x) &= 1 \times qp + 2 \times q^2p + 3 \times q^3p + \cdots = qp(1 + 2q + 3q^2 + \cdots) \\ &= qp(1 - q)^{-2} = qp \times p^{-2} = q/p = (1 - p)/p = p^{-1} - 1 \end{aligned}$$

(D) x is a continuous variable whose probability density function is

$$\begin{aligned} p(x) &= k(6 - x - x^2) \qquad (-3 \leqslant x \leqslant 2) \\ &= 0 \quad \text{elsewhere} \end{aligned}$$

k is a constant. What is the expected value of x?

Solution The constant k must be chosen so that the total probability over the range of x $(-3 \leqslant x \leqslant 2)$ is unity, i.e.

$$\int_{-3}^{2} p(x)\,dx = k\int_{-3}^{2}(6 - x - x^2)\,dx = (125/6)k = 1 \qquad k = 6/125$$

$$E(x) = \int_{-3}^{2} xp(x)\,dx = \frac{6}{125}\int_{-3}^{2} x(6 - x - x^2)\,dx = -1/2$$

EXERCISES 5.2

1. An unbiased die is thrown. What is the expected value of the number thrown? $[3\frac{1}{2}]$
2. 2 balls are drawn, without replacement, from an urn containing 3 black and 4 white balls. The drawer will receive £2.10 for each black ball drawn and £1.40 for each white. What is his expectation? [£3.40]
3. Given that the probability of x successes in n trials is $\binom{n}{x} p^x q^{n-x}$ $(x = 0, 1, 2, \ldots, n)$ where p is the probability of a success and q the probability of a failure in any trial. Show that the mean value of x is np (binomial distribution)
4. The probability density function of a continuous variable x is given by

$$\begin{aligned} p(x) &= ce^{-x/a} \quad (0 \leqslant x \leqslant \infty) \\ &= 0 \text{ elsewhere} \end{aligned}$$

   c is a constant. Find the expected value of x. $[c = 1/a$, expected value $= a]$
5. A chord is drawn, in a random direction, from a point on the circumference of a circle of radius a. Find the mean value of the length of the chord. $[4a/\pi]$

5.7 *The Expectation of Functions of a Random Variable*

The concept of expectation may be generalized. Let $g(x)$ be an arbitrary function of the random variable x, then the expectation of $g(x)$ is defined to be

$$E[g(x)] = \sum_i p_i g(x_i) \quad \text{for a discrete variable} \tag{5.9}$$

$$= \int_{-\infty}^{+\infty} p(x)g(x)\,dx \quad \text{for a continuous variable} \tag{5.10}$$

It follows easily that

$$E[g_1(x) + g_2(x)] = E[g_1(x)] + E[g_2(x)] \tag{5.11}$$

DEFINITION 5.5
The *second moment* of the probability distribution of a random variable x about the origin $(x = 0)$ is defined to be

$$\mu_2' = E(x^2) = \sum_i p_i x_i^2 \quad \text{for a discrete variable} \tag{5.12}$$

$$= \int_{-\infty}^{+\infty} p(x)x^2\,dx \quad \text{for a continuous variable} \tag{5.13}$$

A more important statistical parameter, the *variance* μ_2 of the distribution, is the second moment with respect to the mean μ and is defined to be

$$\mu_2 = E(x-\mu)^2 = \sum_i (x_i-\mu)^2 p_i \quad \text{for a discrete variable} \tag{5.14}$$

$$= \int_{-\infty}^{+\infty} (x-\mu)^2 p(x)\, dx \quad \text{for a continuous variable} \tag{5.15}$$

$$\mu_2 = E(x-\mu)^2 = E(x^2 - 2\mu x + \mu^2)$$

$$= E(x^2) - 2\mu E(x) + \mu^2$$

$$= E(x^2) - \{E(x)\}^2 \quad \text{since } \mu = E(x) \tag{5.16}$$

$$= \mu_2' - \mu^2 \tag{5.17}$$

Definition 5.6
The *standard deviation* σ of a probability distribution is defined by

$$\sigma^2 = \mu_2$$

EXAMPLE 5.7

(A) Show that standard deviation of x in Example 5.6(B) is 104/143 approximately.

Solution

$$\mu_2' = E(x^2) = 0 \times 42/143 + 1^2 \times 72/143 + 2^2 \times 27/143 + 3^2 \times 2/143$$
$$= 198/143$$

Variance $= \mu_2 = \mu_2' - \mu^2 = 198/143 - (132/143)^2 = 10\,890/(143)^2$
Standard deviation $\sigma = \sqrt{\mu_2} = 104/143$ approx.

(B) A discrete variable takes the value x with a probability $e^{-\mu}\mu^x/x!$ ($x = 0, 1, 2, \ldots$). Verify that μ is the value of both the mean and variance of the distribution (Poisson distribution).

Solution

$$\text{Mean} = E(x) = \sum_{x=0}^{\infty} e^{-\mu}\frac{\mu^x}{x!}x = e^{-\mu}\mu\sum_{x=1}^{\infty}\frac{\mu^{x-1}}{(x-1)!} = e^{-\mu}\mu e^{\mu} = \mu$$

Second moment about the origin is

$$\mu_2' = \sum_{x=0}^{\infty} e^{\mu} \frac{\mu^x}{x!} x^2 = \sum_{x=0}^{\infty} \{x(x-1) + x\} e^{-\mu} \frac{\mu^x}{x!} = \sum_{x=0}^{\infty} x(x-1) \frac{e^{-\mu}\mu^x}{x!} + \mu$$

$$= e^{-\mu}\mu^2 \sum_{x=2}^{\infty} \frac{\mu^{x-2}}{(x-2)!} + \mu = e^{-\mu}\mu^2 e^{\mu} + \mu = \mu^2 + \mu$$

$$\text{Variance} = \mu_2 = \mu_2' - \mu^2 = \mu$$

EXERCISES 5.3

1. With reference to Exercise 5.2, no. 1, show that the standard deviation of the probability distribution of throws of an unbiased die is 1.71 approximately.
2. Show the variance of the probability distribution of Exercises 5.2, no. 4 is a^2.
3. Show that for the binomial distribution (see Exercises 5.2, no. 3) the variance is npq.
4. A chord is drawn in a random direction from a point on the circumference of a circle of radius a. Find the mean and the variance of the length of the chord. [$4a/\pi$, $2a^2(1 - 8/\pi^2)$]
5. A point P is selected at random on a line AB of length $2a$. Show that the expected value of the area AP $\times$ PB is $2a^2/3$ and that the probability that the area will exceed $a^2/2$ is $1/\sqrt{2}$.
6. A chord is drawn parallel to a given diameter of a circle of radius a, its distance from the centre of the circle being chosen at random. Show that the mean and variance of the length of the chord are $\pi a/2$ and $a^2(32 - 3\pi^2)/12$ respectively. Show that the probability that the length of the chord will exceed $a\sqrt{3}$ is 1/2.

5.8 *Joint Probability Density Functions*

So far, probability density functions for one random variable only have been considered. However, many problems arise which involve two or more random variables. For simplicity, consider first a case involving two continuous random variables, x and y only.

If a continuous function $p(x, y)$ exists such that the probability of the simultaneous occurrence of particular values of x and y is $p(x, y)$, then $p(x, y)$ is said to be the *joint probability density function* of the two random variables x and y.

The probability that x falls in the interval dx and that y falls simultaneously in the interval dy is therefore $p(x, y)\,dx\,dy$ and it follows that

$$\int_{-\infty}^{+\infty}\int_{-\infty}^{+\infty} p(x, y)\,dx\,dy = 1 \tag{5.18}$$

If x and y are discrete variables, the corresponding result is

$$\sum_i \sum_j p_{ij} = 1 \tag{5.19}$$

where p_{ij} denotes the probability that x and y take simultaneously the values x_i, y_j respectively.

DEFINITION 5.7

Two random variables are said to be *independent* when the probability that x will assume a particular value is independent of the value of y, and conversely.

Two such variables are said to be independently distributed and their probability density functions (p.d.f.) will be functions of x only and y only respectively. Let $p_1(x)$ and $p_2(y)$ denote the respective probability density functions, then

$$p(x, y) = p_1(x)p_2(y) \tag{5.20}$$

The generalization of the above definition for several independent variables is obvious. If $p_i(x_i)$ are the p.d.f.s of n independently distributed random variables x_i $(i = 1, 2, \ldots, n)$ then their joint p.d.f. $p(x_1, x_2, \ldots, x_n)$ is given by

$$p(x_1, x_2, \ldots, x_n) = \prod_i p_i(x_i) \tag{5.21}$$

It follows that, if $\phi(x)$ and $\psi(y)$ are arbitrary functions of *independent* random variables x and y, the expected value of $\phi(x)\psi(y)$ is given by

$$\begin{aligned} E[\phi(x)\psi(y)] &= \iint \phi(x)\psi(y)p(x, y)\, dx\, dy \\ &= \int \phi(x)p_1(x)\, dx \int \psi(y)p_2(y)\, dy \\ &= E[\phi(x)]E[\psi(y)] \end{aligned} \tag{5.22}$$

Generalizing, if x_i $(i = 1, 2, \ldots, n)$ are n independent random variables

$$E\left[\prod_i g_i(x_i)\right] = \prod_i E[g_i(x_i)] \tag{5.23}$$

5.9 Probability Generating Functions

DEFINITION 5.8
The probability generating function $G(t)$ of a random variable x, discrete or continuous, is defined by

$$G(t) = E(t^x) \tag{5.24}$$

When x is a discrete variable, taking the values x_i with probabilities p_i $(i = 1, 2, \ldots, n)$,

$$G(t) = \sum_{i=1}^{n} p_i t^{x_i} \tag{5.25}$$

Assuming that the series in eqn (5.25) may be summed to determine the function $G(t)$, then, by definition, it follows that the probability that x takes the value x_i is the coefficient of t^{x_i} in the expansion of $G(t)$ in powers of t. Thus $G(t)$ may be used to generate the probabilities of the distribution of x and is called the probability generating function (p.g.f.) of x.

If x is a continuous variable

$$G(t) = \int_{-\infty}^{+\infty} t^x p(x)\, dx \tag{5.26}$$

The mean and the variance of a probability distribution may be conveniently derived from $G(t)$. For the discrete variable, put $t = 1$ in (5.25) and we have

$$G(1) = \sum_{i=1}^{n} p_i = 1 \tag{5.27}$$

Differentiating (5.25) with respect to t, we have

$$G'(t) = \sum_{i=1}^{n} p_i x_i t^{x_i - 1} \tag{5.28}$$

$$G'(1) = \sum_{i=1}^{n} p_i x_i = \mu \tag{5.29}$$

Multiplying (5.28) by t, we have

$$tG'(t) = \sum_{i=1}^{n} p_i x_i t^{x_i} \tag{5.30}$$

and differentiating (5.30) with respect to t,

$$G'(t) + tG''(t) = \sum_{i=1}^{n} p_i x_i^2 t^{x_i-1}$$

so that

$$G'(1) + G''(1) = \sum_{i=1}^{n} p_i x_i^2 = E(x^2) = \mu_2' \tag{5.31}$$

Since μ and μ_2' are now known, the variance μ_2 is given by

$$\mu_2 = \sigma^2 = \mu_2' - \mu^2$$

5.10 The Probability Generating Function of the Sum of Independent Random Variables

If x and y are a pair of independent random variables

$$E(t^{x+y}) = E(t^x t^y) = E(t^x)E(t^y)$$

Let the p.g.f.s of the variables x, y and their sum be respectively $G_x(t)$, $G_y(t)$ and $G_{x+y}(t)$ so that

$$G_{x+y}(t) = E(t^{x+y}) = G_x(t)G_y(t) \tag{5.32}$$

Thus the p.g.f. of the sum of two independent random variables equals the product of the p.g.f.s of the individual variables.

It follows that, if $z = \sum_{i=1}^{n} x_i$ $(i = 1, 2, \ldots, n)$ where the x_i are independent random variables, then

$$G_z(t) = \prod_{i=1}^{n} G_{x_i}(t) \tag{5.33}$$

EXAMPLE 5.8

(A) The probability density function of a discrete variable x is $(1/2)^x$ where x takes discrete values 1, 2, 3, . . . but is zero for all other values of x. Obtain the p.g.f. and hence calculate the mean and the variance of the distribution.

Solution

$$G(t) = E(t^x) = \sum_{x=1}^{\infty} (1/2)^x t^x = \sum_{x=1}^{\infty} (t/2)^x = t/(2 - t)$$

Therefore $G'(t) = 2(2 - t)^{-2}$ and $G''(t) = 4(2 - t)^{-3}$

$$\text{Mean, } \mu = G'(1) = 2$$

$$\mu_2' = G'(1) + G''(1) = 2 + 4 = 6$$

$$\text{Variance, } \mu_2 = \mu_2' - \mu^2 = 6 - 2^2 = 2$$

(B) The probability density function of a discrete variable x is given by

$$p(x) = e^{-\mu}\mu^x/x! \quad (x = 0, 1, 2, \ldots)$$

(Poisson distribution: see Example 5.7(B).)
Find the p.g.f. of the variable and deduce that the mean and the variance of the distribution are both μ.

Solution

$$G(t) = \sum_{x=0}^{\infty} e^{-\mu} \frac{\mu^x}{x!} t^x$$

$$= e^{-\mu} \sum_{x=0}^{\infty} (\mu t)^x/x! = e^{-\mu}e^{\mu t} = e^{\mu(t-1)}$$

Therefore

$$G'(t) = \mu e^{\mu(t-1)} \quad \text{and} \quad G''(t) = \mu^2 e^{\mu(t-1)}$$

$$G'(1) = \mu \quad \text{and} \quad G''(1) = \mu^2$$

Thus, Mean $= G'(1) = \mu$

$$\mu_2' = G'(1) + G''(1) = \mu + \mu^2$$

$$\text{Variance } \mu_2 = \sigma^2 = \mu_2' - \mu^2 = \mu$$

(C) An unbiased coin is tossed 5 times. Each head scores 3 and each tail scores -2. What is the probability of scoring more than 5?

Solution Consider this type of problem more generally. Suppose p_i, q_i be the respective probabilities of success and failure in the ith trial of a series of n

independent trials. Let the scores for each success and each failure be respectively α, β.

In the first trial, p_1 is the probability that the score $x_1 = \alpha$ and q_1 that $x_1 = \beta$, so that the p.g.f. $= (p_1t^\alpha + q_1t^\beta)$. For n independent trials, it follows by eqn (5.33) that the p.g.f. of the total score $z = \sum_i x_i$ is

$$\prod_i (p_it^\alpha + q_it^\beta)$$

If the probabilities p_i and q_i are constant (i.e. p and q) from trial to trial, the p.g.f. of the total score in n independent trials is

$$(pt^\alpha + qt^\beta)^n$$

The only possible scores are equal to the powers of t which occur in the expansion of $(pt^\alpha + qt^\beta)^n$.

In the above example, the p.g.f. of the total score is

$$[\tfrac{1}{2}t^3 + \tfrac{1}{2}t^{-2}]^5 = \frac{t^{-10}}{32}(t^5+1)^5$$

$$= \frac{t^{-10}}{32}\{t^{25} + 5t^{20} + 10t^{15} + 10t^{10} + 5t^5 + 1\}$$

$$= \frac{1}{32}\{t^{15} + 5t^{10} + 10t^5 + 10 + 5t^{-5} + t^{-10}\}$$

The only possible scores are 15, 10, 5, 0, -5, -10. The probability of scoring more than 5, i.e. of scoring 10 or 15, is

$$\frac{1}{32}(1+5) = 3/16$$

EXERCISES 5.4

1. With reference to Exercises 5.2, no. 3, show that the p.g.f. for the binomial distribution for n independent trials is $(pt + q)^n$. Deduce that the mean and the variance of the distribution are np and npq respectively.
2. A coin is tossed 4 times. Each head scores 2 and each tail -1. What is the probability of scoring 2 or more? [11/16]
3. An unbiased coin is thrown 5 times. Each time a head is thrown, 2 is scored, and each time a tail appears, 1 is subtracted from the score. Show that the p.g.f. of the score is $(t^3 + 1)^5/(2t)^5$. Hence show that the probability of scoring 1 is 5/16 and of scoring 1 or more is 13/16.

4. A die is rolled 5 times. It is agreed that a throw of 5 or 6 will score 1 and that any other throw will score -1. What is the probability of scoring 1 or more? [17/81]
5. Show that the p.g.f. for the score of an unbiased die is $t(1 - t^6)/6(1 - t)$. Such a die is rolled 5 times; show that the probability of a score of 15 is 0.0837 approximately.
6. If all points of the x-axis between $x = 0$ and $x = a$ are equally probable, show that the p.g.f. of the distribution of x is $(t^a - 1)/a \log_e t$.

5.11 *Binomial and Multinomial Theorems*

The expansion of the binomial expression $(x + y)^n$ can easily be obtained by a simple combinatorial method. Consider the expression in the form of a product

$$(x + y)(x + y) \cdots (x + y) \quad \text{to } n \text{ factors}$$

The general term of the expansion is $Cx^{n-r}y^r$. We wish to find the coefficient C.

The term $x^{n-r}y^r$ arises by selecting x from $(n - r)$ of the factors and y from the r remaining factors. The coefficient C of the general term is the number of ways in which the term $x^{n-r}y^r$ can arise. C is therefore the number of ways in which n factors may be divided into two groups such that one group contains $(n - r)$ factors and the other group contains r factors. Therefore

$$C = \binom{n}{r}$$

Hence

$$(x + y)^n = x^n + \binom{n}{1}x^{n-1}y + \binom{n}{2}x^{n-2}y^2 + \cdots + y^n$$

$$= \sum_{r=0}^{n} \binom{n}{r} x^{n-r}y^r \tag{5.34}$$

By a simple extension of this method, the expansion of the multinomial expression $(x_1 + x_2 + x_3 + \cdots + x_k)^n$ may be obtained.

The general term of the expansion of this expression is

$$Cx_1^{n_1}x_2^{n_2}x_3^{n_3} \cdots x_k^{n_k} \qquad \text{where } n = \sum_{i=1}^{k} n_i$$

The above term arises by selecting x_1 from n_1 of the n factors, x_2 from n_2 of the factors, x_3 from n_3 of the factors, and so on. The number of ways in which a term of this type will arise is equal to the number of ways in which the n factors may be divided into k groups of $n_1, n_2, n_3, \ldots, n_k$ factors. Thus, by eqn (5.6),

$$C = n!/n_1!\, n_2!\, n_3! \cdots n_k!$$

Therefore

$$(x_1 + x_2 + x_3 + \cdots + x_k)^n = \sum \frac{n!}{n_1!\, n_2!\, n_3! \cdots n_k!} x_1^{n_1} x_2^{n_2} x_3^{n_3} \cdots x_k^{n_k} \tag{5.35}$$

subject to $n = \sum_{i=1}^{k} n_i$ and each n_i takes integral values from 0 to n.

5.12 *Independent Trials with Two Outcomes*

Consider a series of repetitive independent experiments for which the outcome of any one experiment is either the occurrence of a certain event or its non-occurrence. The tossing of a coin, for which the outcome is either a head or no head, is an example of such an experiment.

Let p be the probability of the occurrence of such an event so that $(1 - p) = q$ is the probability of its non-occurrence. The occurrence of the event will be called a *success* and the non-occurrence a *failure*. Thus the probabilities of a success and a failure are respectively p and q.

Suppose that n independent trials are made. The probability of x successes, which are necessarily associated with $(n - x)$ failures, in any *one* particular order is $p^x q^{n-x}$.

But x successes and $(n - x)$ failures may occur in $\binom{n}{x}$ different orders in n trials and each permutation occurs with a probability $p^x q^{n-x}$. Since these occurrences are *mutually exclusive* it follows that the total probability of x successes in n trials is given by

$$f(x) = \binom{n}{x} p^x q^{n-x} = \frac{n!}{x!\,(n-x)!} p^x q^{n-x} \tag{5.36}$$

$$(x = 0, 1, 2, \ldots, n)$$

$f(x)$ is known as the Binomial or Bernouilli* Probability Function. It should be noted that the probabilities in (5.36) are generated as terms of

* Bernouilli was a pioneer in applying probability theory to discrete variables.

the binomial expansion of

$$(q + p)^n = q^n + \binom{n}{1} q^{n-1} p + \binom{n}{2} q^{n-2} p^2 + \cdots + \binom{n}{x} q^{n-x} p^x + \cdots + p^n = \sum_{x=0}^{n} f(x) \tag{5.37}$$

Note that, as expected, $\sum_{x=0}^{n} f(x) = 1$ since $(p + q) = 1$.

Suppose that N samples, each of size n, are drawn from a large statistical population. Since, in the long run, probabilities may be taken to be relative frequencies, it follows that, if N is large enough, the frequencies of 0, 1, 2, . . . , n successes, occurring in N samples, will be the successive terms of the expansion of $N(q + p)^n$.

5.13 The Probability Generating Function of the Binomial Distribution

From eqns (5.25) and (5.36), the p.g.f. of the binomial distribution is given by

$$G(t) = \sum_{x=0}^{n} \binom{n}{x} p^x q^{n-x} t^x = \sum_{x=0}^{n} \binom{n}{x} (pt)^x q^{n-x} = (q + pt)^n \tag{5.38}$$

$$G'(t) = np(q + pt)^{n-1} \quad \text{and} \quad G''(t) = n(n-1)p^2(q + pt)^{n-2}$$

$$G'(1) = np \quad \text{and} \quad G''(1) = n(n-1)p^2$$

Therefore, by (5.29), Mean, $\mu = G'(1) = np$ (5.39)

and by (5.31), $\mu_2' = G'(1) + G''(1) = np\{1 + (n - 1)p\}$

$$\text{Variance, } \mu_2 = \mu_2' - \mu^2 = np\{1 + (n-1)p - np\} = np(1 - p) = npq \tag{5.40}$$

$$\text{Standard deviation, } \sigma = \sqrt{(npq)} \tag{5.41}$$

EXAMPLE 5.9
Expand $(x + 2y - 3z)^3$.

Solution Consider first the expansion of $(x + y + z)^3$. The expansion may be expressed as

$$\sum_{n_1, n_2, n_3} \frac{3!}{n_1!\, n_2!\, n_3!} x^{n_1} y^{n_2} z^{n_3}$$

where $n_1 + n_2 + n_3 = 3$.

The following is a list of all possible partitions of 3: (3, 0, 0) (2, 1, 0) (2, 0, 1) (1, 1, 1) (1, 0, 2) (1, 2, 0) (0, 2, 1) (0, 1, 2) (0, 3, 0) (0, 0, 3). Therefore

$$(x + y + z)^3 = x^3 + y^3 + z^3 + 3x^2y + 3xy^2 + 3y^2z + 3yz^2 + 3x^2z + 3xz^2 + 6xyz$$

$$(x + 2y - 3z)^3 = x^3 + 8y^3 - 27z^3 + 6x^2y + 12xy^2 - 36y^2z + 54yz^2 - 9x^2z + 27xz^2 - 36xyz$$

EXAMPLE 5.10
10 dice are thrown and a throw of 5 or 6 is a "success". Find the probability of (*a*) 3 successes, (*b*) 3 successes at most, (*c*) 3 successes or more.

Solution The probability of a success = probability of throwing 5 or 6

$$= p = 1/3$$

Therefore $q = 2/3$.

The probability of x successes in a throw of 10 dice (or in 10 throws of one die) is

$$\binom{10}{x} p^x q^{10-x} = \binom{10}{x} \left(\frac{1}{3}\right)^x \left(\frac{2}{3}\right)^{10-x}$$

(*a*) The probability of 3 successes is $\binom{10}{3}\left(\frac{1}{3}\right)^3\left(\frac{2}{3}\right)^7 = 0.260$ approx.

(*b*) The probability of 0, 1 and 2 successes is

$$\left(\frac{2}{3}\right)^{10} + \binom{10}{1}\left(\frac{1}{3}\right)^1\left(\frac{2}{3}\right)^9 + \binom{10}{2}\left(\frac{1}{3}\right)^2\left(\frac{2}{3}\right)^8$$

$$= 0.0173 + 0.0867 + 0.1951 = 0.299 \text{ approx.}$$

Thus, probability of 3 successes at most is

$$0.260 + 0.299 = 0.559 \text{ approx.}$$

(*c*) The probability of 3 successes or more is

$$1 - (\text{probability of 0, 1 and 2 successes}) = 1 - 0.299 = 0.701 \text{ approx.}$$

EXAMPLE 5.11

In a precision bombing attack, there is a 50% chance that any one bomb will strike the target. Two direct hits are required to destroy the target completely. How many bombs must be dropped to give approximately 99% chance of destroying the target?

Solution The probability that a bomb will strike the target is $p = 1/2$. Thus $q = 1/2$. Let n bombs be dropped. Two or more hits will destroy the target. The probability of 2 or more hits from n bombs is

$$1 - (\text{the sum of the probabilities of 0 and 1 hits from } n \text{ bombs})$$

$$= 1 - \left\{\left(\frac{1}{2}\right)^n + \binom{n}{1}\left(\frac{1}{2}\right)\left(\frac{1}{2}\right)^{n-1}\right\} = 1 - (n+1)(1/2)^n$$

The minimum number of bombs is given by the least value of n such that

$$1 - (n+1)(1/2)^n \geqslant 0.99$$

$$\text{or } (n+1)(1/2)^n \leqslant 0.01$$

For $n = 11$, $12(1/2)^{11} = 0.0058 < 0.01$

For $n = 12$, $13(1/2)^{12} = 0.0318 > 0.01$

Therefore, the least number of bombs required is 11.

EXAMPLE 5.12

00 samples, each containing 20 components, produced by a machine are tested for defectives. The following table gives the frequency distribution of samples containing 0, 1, 2, . . . defectives.

No. of defectives per sample	0	1	2	3	4	5	6	7 or more
No. of Samples	21	35	25	14	4	1	0	0

Show that the mean number of defectives per sample is 1.48 and hence that an estimate of p, the probable proportion of defectives in the population, is 0.074. Assuming a binomial distribution, use this value of p to calculate the probable number of samples containing 0, 1, 2, . . . , 6 defectives.

Solution

$$\begin{aligned}\text{Mean no. of defectives per sample} &= \tfrac{1}{100}(21 \times 0 + 1 \times 35 + 2 \times 25 + 3 \times 14 \\ &\quad + 4 \times 4 + 5 \times 1 + 6 \times 0 + 7 \times 0) \\ &= 148/100 = 1.48\end{aligned}$$

Proportion of defectives in the population $= p = \frac{1}{20}(1.48) = 0.074$. Thus $q = 0.926$.

On the assumption of a binomial distribution, the probable number of samples containing x defectives is $100\binom{20}{x}p^x q^{20-x}$.

Thus the probable number of samples containing 0, 1, 2, . . . defectives is

$$\begin{aligned}&100\{q^{20}, 20pq^{19}, 190p^2q^{18}, \ldots\} \\ &= 100\{(0.926)^{20}, 20(0.074)(0.926)^{19}, 190(0.074)^2(0.926)^{18}, \ldots\} \\ &= 21, 34, 26, 12, 4, 1, 0\end{aligned}$$

5.14 Combinatorial Generating Functions

The calculation of the number of outcomes favourable to a certain event is often difficult. In Section 5.9, the concept of probability generating functions was introduced and some applications to problems of enumeration given. In this section, some examples of the application of other generating functions to problems of enumeration associated with the multinomial theorem (eqn 5.35) will be given.

Consider the problem of Example 5.5 (p. 131) concerning the throwing of 6 balls into 4 boxes. It will now be shown that the multinomial expansion

$$(x_1 + x_2 + x_3 + x_4)^6 = \sum_{\substack{n_1, n_2, \\ n_3, n_4}} \frac{6!}{\prod_{i=1}^{4} n_i!} x_1^{n_1} x_2^{n_2} x_3^{n_3} x_4^{n_4} \tag{5.42}$$

$$(n_1 + n_2 + n_3 + n_4 = 6)$$

may be interpreted in terms of the above problem. It will be noted that the coefficient of the general term

$$\frac{6!}{\prod_{i=1}^{4} n_i!} x_1^{n_1} x_2^{n_2} x_3^{n_3} x_4^{n_4} \tag{5.43}$$

gives the number of ways in which 6 balls may be partitioned so that n_1, n_2, n_3 and n_4 balls fall into 4 boxes respectively. On putting $x_1 = x_2 = x_3 = x_4 = 1$ in (5.42), it follows that the total number of possible outcomes is

$$\sum_{\substack{n_1, n_2, \\ n_3, n_4}} \frac{6!}{\prod_{i=1}^{4} n_i!} = 4^6$$

From eqn (5.43), $6!/n_1!\, n_2!\, n_3!\, 2!$ gives the number of outcomes for which 2 balls fall into the 4th box whilst n_1, n_2, n_3 balls ($n_1 + n_2 + n_3 = 4$) fall respectively into the other 3 boxes. Therefore the number of outcomes for which 2 balls fall into the 4th box is

$$\frac{6!}{2!\, 4!} \sum_{n_1, n_2, n_3} \frac{4!}{(n_1!\, n_2!\, n_3!)}$$

$$= \binom{6}{2}(x_1 + x_2 + x_3)^4 = \binom{6}{2} 3^4 \qquad (x_1 = x_2 = x_3 = 1)$$

$$\text{Required probability} = \binom{6}{2} 3^4 \Big/ 4^6 \qquad \text{(as before)}$$

The expression $(x_1 + x_2 + x_3 + x_4)^6$ provides a simple generating function for the solution of many similar problems concerning this particular physical system. This g.f. is of little value for the problem considered since it can be solved more simply otherwise. Nevertheless, the treatment is illustrative of the application of generating functions to much more complex problems when the enumeration of outcomes is otherwise difficult.

We now introduce another type of generating function by applying it to the following problem. An urn contains 3 black balls and 5 white. 3 balls are successively drawn at random from the urn and are placed in a black box. The remaining 5 balls are placed in a white box. Find the probability that the sum of the number of black balls in the black box and the number of white balls in the white box is 4.

This problem may be solved easily without a generating function. The total number of ways in which 3 balls may be drawn from the urn is $\binom{8}{3}$. The only way in which 4 may be obtained as the sum of the number of black balls in the black box and of white balls in the white box is when there are 3 white balls in the white box and 1 black ball in the black box. The black box will then contain 1 black ball and 2 white. The number of ways of filling the black box will therefore be $\binom{3}{1}\binom{5}{2}$ which is the number of outcomes favourable to the total of 4. Therefore

$$\text{Required probability} = \binom{3}{1}\binom{5}{2}\Big/\binom{8}{3} = 15/28$$

Now consider the generating function

$$(x_1t + x_2)^3(x_1 + x_2t)^5 \tag{5.44}$$

where, similarly to the last example, the variables x_1, x_2 are associated with the two boxes: x_1 with the black and x_2 with the white. In eqn (5.44), the first factor is concerned with the partitioning of the 3 black balls between the two boxes and the second factor with the partitioning of the 5 white balls between them.

In this problem, we are interested only in those outcomes for which 3 balls are in the black box and 5 are in the white box so that we are concerned with the enumerations which occur as coefficients of $x_1^3x_2^5$ in the expansion of (5.44). Since t is associated with x_1 in the first factor and with x_2 in the second factor of (5.44), the coefficient of $x_1^3x_2^5t^r$ in the expansion will give the number of ways in which a total of r balls with colours matching the boxes occur amongst all the possible outcomes for which 3 balls are in the black box and 5 balls in the white box.

Putting $t = 1$ in (5.44), the g.f. becomes $(x_1 + x_2)^8$ from which the total number of possible outcomes is

$$\text{Coefficient of } x_1^3x_2^5 \text{ in the expansion of } (x_1 + x_2)^8 = \binom{8}{3} \text{ as before}$$

Equation (5.44) may be written in the form

$$\left\{\sum_{m=0}^{3}\binom{3}{m}(x_1t)^{3-m}x_2^{\,m}\right\}\left\{\sum_{n=0}^{5}\binom{5}{n}x_1^{5-n}(x_2t)^n\right\}$$

so that the general term of the product is

$$\binom{3}{m}\binom{5}{n}x_1^{8-(m+n)}x_2^{(m+n)}t^{3-(m-n)}$$

The term containing $x_1{}^3x_2{}^5t^4$ is that for which $m = 2$, $n = 3$ and is

$$\binom{3}{2}\binom{5}{3}x_1{}^3x_2{}^5t^4$$

so that the number of outcomes favourable to the sum of matching colours equalling 4 is

$$\binom{3}{2}\binom{5}{3} = \binom{3}{1}\binom{5}{2} \quad \text{as before}$$

The generating function method has again been illustrated by application to a problem which can be solved much more easily otherwise. Nevertheless, such a problem would become much more complicated by the introduction of even one more colour and the g.f. method would be quite valuable.

Consider the following problem: an urn contains n balls, n_1 being black, n_2 being white and n_3 red. If m_1 balls are drawn and are placed in a black box, m_2 are drawn and placed in a white box, and the remaining m_3 are placed in a red box, find the probability that a total of r balls have the colours of the boxes in which they are placed.

The appropriate generating function is

$$(x_1t + x_2 + x_3)^{n_1}(x_1 + x_2t + x_3)^{n_2}(x_1 + x_2 + x_3t)^{n_3}$$

when $n = n_1 + n_2 + n_3 = m_1 + m_2 + m_3$.

The number of outcomes favourable to the desired event is the coefficient of $x_1{}^{m_1}x_2{}^{m_2}x_3{}^{m_3}t^r$. It is much more difficult to find the number of favourable outcomes without the use of the above g.f.

EXERCISES 5.5

1. (*a*) Find the coefficient of x^3yz^2 in the expansion of $(2x - 3y + 5z)^6$.

   (*b*) Find the expansion of $(a - 2b - 3c)^4$.

   [(*a*) $-36\,000$

   (*b*) There are 15 terms. The expansion is $a^4 - 8a^3b - 12a^3c + 24a^2b^2 + 54a^2c^2 + 72a^2bc - 32ab^3 - 108ac^3 - 216abc^2 - 144ab^2c + 16b^4 + 96b^3c + 216bc^3 + 216b^2c^2 + 81c^4$]

2. A die is thrown 10 times. Show that the probability that a six appears at least twice is 0.516 approximately.
3. In a factory, 10% of the output of a certain component is defective.
   (*a*) What is the probability that two or more components will be defective in a random sample of 10?
   (*b*) What is the largest sample size which will be at least 50% certain to contain no defectives. [(*a*) 0.265 approx., (*b*) 6]
4. What is the probability that there are not more than 3 boys in a family of 8 children? (Assume that the probability of a male or female birth is $\frac{1}{2}$.) [93/256]
5. A machine produces parts of which 1% are defective. What is the smallest sample size which should be inspected in order that the probability that there will be no defectives in the sample is less than 0.05? [296]
6. On the average, hens lay eggs 4 days per week. On how many days during a season of 200 days may a poultry keeper expect to obtain 6 eggs from 8 hens? If he obtained 6 eggs or more on 42 days during the season, would this suggest that factors other than chance were operating? [36 approx.; yes]
7. If $f(x)$ is the probability of x successes in a binomial distribution [see eqn (5.36)], show that

$$f(x+1) = \frac{p}{q}\left(\frac{n-x}{x+1}\right) f(x)$$

8. 6 balls are thrown into 3 boxes and each ball is equally likely to fall into either box.
   (*a*) Show that the probability that 3 balls fall into the second box is 160/729
   (*b*) Show that the probability that each box will contain at least one ball is 20/27
9. An urn contains 11 balls of which 6 are black and 5 are white. 4 balls are drawn at random from the urn and are placed in a black box. The remaining 7 balls are placed in a white box. Using a generating function, verify that the probability that a total of 5 balls match the colours of the boxes containing them is $\binom{5}{2}\binom{6}{2}\Big/\binom{11}{4}$.

5.15 Compound Probability

Many applications of probability are concerned with the joint occurrence of two events A and B associated with an experiment.

In the Venn diagrams of Fig. 5.5, let the rectangles denote the sample space for the experiment and let the sets A and B comprise those sample points associated with the occurrence of the events A and B respectively.

The set $A \cup B$ of sample points (Fig. 5.5(*c*)) comprises all those sample points which are associated with the occurrence of the events A or B or both A and B. It follows that the probability of the occurrence of the event

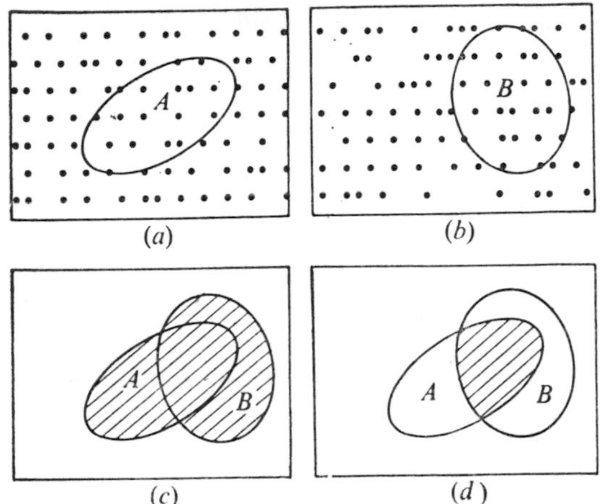

Figure 5.5
(a) Set A of sample points associated with event A
(b) Set B of sample points associated with event B
(c) Set $A \cup B$ (d) Set $A \cap B$

or the event B or both, denoted by $p(A \cup B)$, will be the sum of the robabilities associated with all the sample points in the set $A \cup B$.

Similarly the set $A \cap B$ (Fig. 5.5(*d*)), comprising the sample points which re common to the sets A and B, contains those sample points which are ssociated with the simultaneous occurrence of the events A and B. The robability of the simultaneous occurrence of A and B will be denoted by $(A \cap B)$.

Since $A \cup B = A + B - (A \cap B)$ it follows that

$$p(A \cup B) = p(A) + p(B) - p(A \cap B) \tag{5.45}$$

If the two events A and B are mutually exclusive, the occurrence of one vent precludes the occurrence of the other which implies that the sets A and have no sample points in common, i.e. $A \cap B = \varnothing$.

Therefore, for two *mutually exclusive* events A and B,

$$p(A \cup B) = p(A) + p(B) \tag{5.46}$$

t may easily be verified, from a Venn diagram for three sets of sample oints associated with events A, B and C, that the probability of A or B r C is

$$p(A \cup B \cup C) = p(A) + p(B) + p(C) - p(A \cap B) - p(B \cap C) - p(C \cap A) + p(A \cap B \cap C) \tag{5.47}$$

In general, for n events $A_1, A_2, \ldots, A_n$, it may be shown that

$$p(A_1 \cup A_2 \cup \cdots \cup A_n) = \sum_{i=1}^{n} p(A_i) - \sum_{\substack{i,j \\ i<j}} p(A_i \cap A_j) + \sum_{\substack{i,j,k \\ i<j<k}} p(A_i \cap A_j \cap A_k) - \cdots (-1)^{n-1} p(A_1 \cap A_2 \cap \cdots \cap A_n) \quad (5.48)$$

If the n events are mutually exclusive, all terms on the right-hand side of (5.48) except the first, are zero. Therefore

$$p(A_1 \cup A_2 \cup \cdots \cup A_n) = \sum_{i=1}^{n} p(A_i) \quad (5.49)$$

for n mutually exclusive events.

5.16 *Conditional Probability*

Suppose that an event A has occurred and that we now wish to find the probability that an event B will occur subject to this condition. This is a *conditional probability*.

With reference to Fig. 5.5, since the event A has occurred, we are now concerned only with the set A of sample points associated with that event. Thus the set A now comprises the whole sample space from which the conditional probability of the event B is to be determined. Since the total probability residing in any complete sample space must be 1, it will be necessary to multiply the probabilities p_i originally assigned to the sample points of the set A by a factor k so that the sum of the new probabilities kp_i of the sample points of set A shall be 1.

Thus k is determined by the condition

$$\sum_{A} kp_i = 1$$

Therefore

$$k = 1 \Big/ \sum_{A} p_i = 1/p(A) \quad \text{provided } p(A) \neq 0$$

The sample points of the set A will therefore be assigned probabilities $p_i/p(A)$.

Let $p(B/A)$ denote the conditional probability of the event B, i.e. the probability that B will occur when A has already occurred. The measure of $p(B/A)$ will thus be the sum of the probabilities $p_i/p(A)$ assigned to those sample points associated with the occurrence of the event B which are in the set A. Therefore

$$p(B/A) = \sum_{A \cap B} p_i/p(A) = \frac{\sum_{A \cap B} p_i}{p(A)} = \frac{p(A \cap B)}{p(A)} \tag{5.50}$$

$$p(A \cap B) = p(A)p(B/A) \tag{5.51}$$

By interchanging the order of the events, we have

$$p(A \cap B) = p(B)p(A/B) \tag{5.52}$$

Since $p(A \cap B)$ is the probability of the simultaneous occurrence of events A and B, eqn (5.51) is the formula by which this probability may be determined when the events A and B are *not* independent. The formula (5.51) may easily be generalized for several independent events. For three such events A, B and C, we have

$$\begin{aligned} p(A \cap B \cap C) &= p(A \cap B)p(C/A \cap B) \\ &= p(A)p(B/A)p(C/A \cap B) \end{aligned} \tag{5.53}$$

and similar results may be obtained by permuting the symbols A, B and C. By induction, it easily follows that, for n dependent events $A_1, A_2, A_3, \ldots, A_n$,

$$\begin{aligned} p(A_1 \cap A_2 \cdots \cap A_n) = p(A_1)p(A_2/A_1)&p(A_3/A_1 \cap A_2) \\ &\times p(A_4/A_1 \cap A_2 \cap A_3) \cdots \\ &\times p(A_n/A_1 \cap A_2 \cap \cdots \cap A_{n-1}) \end{aligned} \tag{5.54}$$

Clearly $n!$ similar equations result from permutations of the symbols $A_1, A_2, \ldots, A_n$.

Now suppose that an event A is *independent* of an event B in the sense that the probability of the occurrence of the event B has the same value whether A occurs or does not, so that

$$p(B/A) = p(B)$$

With reference to (5.51), it follows that, for two independent events,

$$p(A \cap B) = p(A)p(B) \tag{5.55}$$

By comparison with (5.52), it follows that

$$p(A/B) = p(A)$$

which implies that the event A is also independent of the event B. Thus if B is independent of A, A is also independent of B.

From (5.55), it follows that, for three independent events A, B and C

$$p(A \cap B \cap C) = p(A \cap B)p(C) = p(A)p(B)p(C) \tag{5.56}$$

For n independent events, we have the obvious generalization

$$p(A_1 \cap A_2 \cap \cdots \cap A_n) = \prod_{i=1}^{n} p(A_i) \tag{5.57}$$

EXAMPLE 5.13

An urn contains 3 white and 7 black balls. What is the probability that (*a*) one ball drawn is white, (*b*) two drawn are both white, (*c*) two drawn will be of the same colour, (*d*) two drawn will be of different colours, (*e*) two drawn will be 1 white and then 1 black? (The balls are drawn successively without replacement.)

Solution There are altogether 10 balls.

(*a*) The probability of drawing 1 white ball (event A) is 3/10, $p(A) = 3/10$.

(*b*) After drawing 1 white, 2 white and 7 black balls remain.
Thus the probability of drawing a second white (event B) is 2/9.
This is a conditional probability expressed by $p(B/A) = 2/9$.
Therefore the probability of drawing two whites is

$$p(A \cap B) = p(A)p(B/A) = \frac{3}{10} \times \frac{2}{9} = \frac{1}{15}$$

This result may be obtained otherwise. The number of ways of selecting 2 balls from 10 is $\binom{10}{2}$ and the number of ways of selecting 2 white balls from 3 is $\binom{3}{2}$.

Thus, the probability of selecting two white balls $= \binom{3}{2} \Big/ \binom{10}{2} = \frac{1}{15}$.

(*c*) Similarly the probability of drawing two black balls is

$$\frac{7}{10} \times \frac{6}{9} = \frac{7}{15} \quad \text{or (alternatively)} \quad \binom{7}{2} \Big/ \binom{10}{2}$$

Since the drawing of two whites and of two blacks are mutually exclusive events, the probability of drawing *either* two whites *or* two blacks is

$$\frac{1}{15}+\frac{7}{15}=\frac{8}{15}$$

(*d*) The drawing of two balls of different colours is the alternative to (*c*).
The probability of drawing two balls of different colours is $1-\frac{8}{15}=\frac{7}{15}$.
Alternatively, the required probability is $\binom{3}{1}\binom{7}{1}\Big/\binom{10}{2}=\frac{7}{15}$.

(*e*) The probability of drawing firstly a white ball is 3/10.
The conditional probability of then drawing a black is 7/9.

$$\text{Required probability}=\frac{3}{10}\times\frac{7}{9}=\frac{7}{30}$$

As expected, this probability is half that obtained in (*d*).

EXAMPLE 5.14
What is the probability of throwing a sum of 9 and/or a difference of 3 in a single throw of two dice? If a difference of 3 is thrown, what is the probability that the sum is 9?

Solution Let the event A be a throw of total 9 and the event B be a throw of difference 3.
4 outcomes (6, 3) (5, 4) (4, 5) (3, 6) comprise event A. Thus $p(A)=4/36$.
6 outcomes (6, 3) (5, 2) (4, 1) (1, 4) (2, 5) (3, 6) comprise event B. Thus $p(B)=6/36$.
Events A and B are not mutually exclusive since A and B have common elements (6, 3) and (3, 6) so that

$$p(A\cap B)=2/36$$

The required probability of A or B or both is

$$p(A\cup B)=p(A)+p(B)-p(A\cap B)=\frac{4}{36}+\frac{6}{36}-\frac{2}{36}=\frac{2}{9}$$

This result also follows from first principles by noting that there are 8 different outcomes only giving a sum of 9 and/or a difference of 3.
Since $p(A\cap B)\neq p(A)p(B)$, A and B are not independent, and

$$p(A/B)=\frac{p(A\cap B)}{p(B)}=\frac{2/36}{6/36}=\frac{1}{3}\neq p(A)$$

This result again follows easily from first principles. Note that when the event B has occurred, there are 6 sample points only to be considered when calculating $p(A/B)$. Only 2 of these 6, namely (6, 3) and (3, 6), lead to the event A. Therefore $p(A/B) = \frac{2}{6} = \frac{1}{3}$.

EXAMPLE 5.15

Six cards are drawn from a pack. What is the probability that exactly two aces and two kings will be drawn (*a*) if the cards are drawn successively, (*b*) if each card is returned to the pack which is shuffled before the next card is drawn?

Solution

(*a*) Let the aces be denoted by A, the kings by K, and all other cards by X. Firstly, let us calculate the probability that the 6 cards are drawn in the particular order A, A, K, K, X, X.

By eqn (5.54) for $n = 6$, this probability is given by

$$p(A \cap A \cap K \cap K \cap X \cap X)$$
$$= p(A)p(A/A)p(K/A \cap A) \cdots p(X/A \cap A \cap K \cap K \cap X)$$

The probability $p(A)$ of drawing the first ace from a pack of 52 cards is 4/52. When one ace has been drawn, the probability $p(A/A)$ of drawing a second ace from the remaining 51 cards is 3/51 and so on. Therefore

$$p(A \cap A \cap K \cap K \cap X \cap X) = \frac{4}{52} \times \frac{3}{51} \times \frac{4}{50} \times \frac{3}{49} \times \frac{44}{48} \times \frac{43}{47}$$

This is the probability of drawing the 6 cards in one specified order. Since however, there are $6!/(2!)^3$, i.e. 90 permutations of the symbols A, A, K, K, X, X, the required probability is the sum of the probabilities of all these permutations. It is easily seen, however, that the probabilities of each of these permutations are equal, e.g.

$$p(K \cap A \cap A \cap X \cap K \cap X) = \frac{4}{52} \times \frac{4}{51} \times \frac{3}{50} \times \frac{44}{49} \times \frac{3}{48} \times \frac{43}{47}$$
$$= p(A \cap A \cap K \cap K \cap X \cap X)$$

$$\text{Required probability} = 90 \times \frac{4}{52} \times \frac{3}{51} \times \frac{4}{50} \times \frac{3}{49} \times \frac{44}{48} \times \frac{43}{47}$$
$$= 0.0017 \text{ approx.}$$

(*b*) If, after drawing a card, it is returned to the pack which is shuffled before the next card is drawn, the 6 constituent events comprising the compound event become independent so that

$$\text{Required probability} = 90\left(\frac{4}{52}\right)^2\left(\frac{4}{52}\right)^2\left(\frac{44}{52}\right)^2 = 0.0023 \text{ approx.}$$

EXAMPLE 5.16

Three persons A, B, C, in that order, successively throw a coin and the first person to throw a tail wins. What are the respective probabilities that A, B and C will win?

Solution The probability that A will win on his first throw is $\frac{1}{2}$.

The probability that A, B and C all fail to win on their first throws is $(\frac{1}{2})^3$.

Thus the probability that A will win on his second throw is $(\frac{1}{2})(\frac{1}{2})^3$ and the probability that A will win on his third throw is $(\frac{1}{2})(\frac{1}{2})^3(\frac{1}{2})^3$ and so on.

Therefore the total probability that A will win is

$$\tfrac{1}{2} + (\tfrac{1}{2})^4 + (\tfrac{1}{2})^7 + \cdots = \frac{\frac{1}{2}}{1 - (\frac{1}{2})^3} = \tfrac{4}{7}$$

The probabilities that B will win on his first, second, third, . . . , throws are respectively

$$(\tfrac{1}{2})^2,\ (\tfrac{1}{2})^2(\tfrac{1}{2})^3,\ (\tfrac{1}{2})^2(\tfrac{1}{2})^3(\tfrac{1}{2})^3, \ldots$$

Therefore the total probability that B will win is

$$(\tfrac{1}{2})^2 + (\tfrac{1}{2})^5 + (\tfrac{1}{2})^8 + \cdots = \tfrac{2}{7}$$

Similarly, the total probability that C will win is

$$(\tfrac{1}{2})^3 + (\tfrac{1}{2})^6 + (\tfrac{1}{2})^9 + \cdots = \tfrac{1}{7}$$

As expected, the sum of these probabilities is 1.

EXAMPLE 5.17

Eight cards are drawn from a pack one at a time with replacement. Find the probability that these 8 cards will include at least one card from each suit.

Solution Let the event A be that for which the drawing of 8 cards include at least one card from each suit, and let the event B be the alternative for which the drawing of 8 cards lacks at least one of the suits. Therefore

$$p(A) = 1 - p(B)$$

To solve this problem, we shall calculate $p(B)$. All outcomes favourable to the event B may be divided into subsets B_1, B_2, B_3 and B_4 comprising those sets of outcomes which respectively lack hearts, diamonds, clubs and spades. Thus, B_1, B_2, B_3 and B_4 are not mutually exclusive since, for example, an outcome comprising clubs and spades belongs to both B_1 and B_2. By eqn (5.48)

$$p(B) = p(B_1 \cup B_2 \cup B_3 \cup B_4)$$

$$= \sum_i p(B_i) - \sum_{i,j} p(B_i \cap B_j)$$

$$+ \sum_{i,j,k} p(B_i \cap B_j \cap B_k) - p(B_1 \cap B_2 \cap B_3 \cap B_4)$$

Now $p(B_1)$ is the probability that no hearts will appear in a drawing of 8 cards. Therefore $p(B_1) = (3/4)^8$.

Similarly $p(B_2) = p(B_3) = p(B_4) = (3/4)^8$. Therefore

$$\sum_i p(B_i) = 4(\tfrac{3}{4})^8$$

$p(B_1 \cap B_2)$ is the probability that neither hearts nor diamonds appear in a drawing of 8 cards. Therefore $p(B_1 \cap B_2) = (1/2)^8$.

From the four sub-sets B_i, there are $\binom{4}{2}$, i.e. 6, pairs which may be selected, the probability for each pair being also $(1/2)^8$. Therefore

$$\sum_{i,j} p(B_i \cap B_j) = 6(\tfrac{1}{2})^8$$

Similarly

$$\sum_{i,j,k} p(B_i \cap B_j \cap B_k) = 4(\tfrac{1}{4})^8$$

and $p(B_1 \cap B_2 \cap B_3 \cap B_4) = 0$ since the absence of each suit simultaneously is impossible. Therefore

$$p(A) = 1 - 4(\tfrac{3}{4})^8 + 6(\tfrac{1}{2})^8 - 4(\tfrac{1}{4})^8 = 0.623 \text{ approx.}$$

Now consider the additional problem as follows. Cards are drawn successively with replacement, until all suits appear at least once. Find the probability that 8 cards must be drawn.

Let p_n denote the probability that all suits will appear at least once when n cards are drawn. Therefore, as above,

$$p_n = 1 - 4(\tfrac{3}{4})^n + 6(\tfrac{1}{2})^n - 4(\tfrac{1}{4})^n$$

Let p_n' denote the probability that all suits will first appear at the drawing of the nth card. When n cards are drawn, there are probabilities p_4', p_5', p_6', . . . , p_n' that all suits will appear at least once at the 4th, 5th, 6th, . . . , nth drawing and since these outcomes are mutually exclusive

$$p_n = p_4' + p_5' + p_6' + \cdots + p_n' \qquad \text{Thus } p_n' = p_n - p_{n-1}$$

In the given example, $n = 8$. Therefore

$$p_8' = p_8 - p_7 = \{1 - 4(\tfrac{3}{4})^8 + 6(\tfrac{1}{2})^8 - 4(\tfrac{1}{4})^8\} - \{1 - 4(\tfrac{3}{4})^7 + 6(\tfrac{1}{2})^7 - 4(\tfrac{1}{4})^7\} = 0.110$$

EXERCISES 5.6

1. A coin is tossed 3 times. Find the probability that (*a*) the first two tosses include a head and a tail, (*b*) there are more tails than heads. [1/2, 1/2]
2. In a throw of two dice, what is the probability of either a total of 4 or a total of 8? What is the probability of a difference of 4 and/or a total of 8? [2/9, 7/36]
3. 10 persons are seated at a circular table. Find the probability that a particular pair of persons are seated next to each other. [2/9]
4. A die is thrown twice. What is the probability that the sum of the two throws exceeds 10 given that (*a*) one is a 6, (*b*) the first throw is a 6? [3/11, 1/3]
5. A set of 19 cards are numbered 1, 2, 3, . . . , 19. If one card is drawn, what is the probability that its number is a multiple of 2 or 3? What is the probability that the number will be a multiple of 2 or 3 or both? [9/19, 6/19, 12/19]
6. A person holds two tickets for a draw in which there are 12 horses and 15 blanks. Show that the probability of drawing a horse is 82/117.
7. Each of a pair of dice has two faces numbered 1, two numbered 4 and the remaining two numbered 2 and 3. Set up a two dimensional sample space for a single throw of the two dice and indicate the sets of points S_1, S_2 and S_3 corresponding respectively to the events: (*a*) a total score of 6, (*b*) a total score of 4, (*c*) the same score on each die.

   Calculate the probability of (*d*) a score of 4 or 6, (*e*) scoring 4 or 6 when both dice show the same score. [5/18, 1/18]
8. A group of 10 men and 6 women includes one married couple. A committee of 4 men and 3 women is chosen by lottery. What is the probability that both the husband and wife will be members? [1/5]

9. An unbiased die, having 6 sides, is thrown 5 times. Show that the probability that at least one 6 is thrown is 0.598 approximately.

10. A bag contains 4 red balls and 3 blue. Two drawings of 2 balls are made. What is the probability that 2 red balls and then 2 blue balls are drawn if (*a*) the balls are returned to the bag after the first draw, (*b*) the balls are not returned? [2/49, 3/35]

11. In a game of bridge, one player's hand contains K, Q, J, 10, 9 of spades. What is the probability that another player's hand holds the ace of spades and exactly two other spades? [0.115 approx.]

12. Find the probability of obtaining (*a*) more than 7 with a throw of 2 dice, (*b*) at least one 6 in a throw of 3 dice. [5/12, 91/216]

13. From a bag containing 4 white and 5 black balls, 3 are successively drawn at random. What are the odds against all 3 being black? [37 to 5]

14. Two bags contain respectively 3 white and 2 black balls and 5 white and 3 black balls. If a bag is chosen at random and 1 ball is drawn from it, what is the probability that it is white? [49/80]

15. Four screws are selected from a large batch of screws of which 15% are defective. Show that the probability is 0.05% that these screws are all defective.

16. Three groups, each of 5 people, comprise 3 men and 2 women, 2 men and 3 women and 1 man and 4 women. One person is selected at random from each group. Show that probability is 58/125 that the selected group comprises 1 man and 2 women.

17. Two drawings, each of 3 balls, are made from a bag containing 5 white balls and 8 black. Find the probability that the first drawing gives 3 white balls and is followed by a second drawing of 3 black balls, the balls being (*a*) replaced, (*b*) not replaced before the second trial. [140/20 449, 7/429]

18. A bridge player and his partner hold 9 hearts between them. What is the probability that their opponents hold 2 hearts each? [234/575]

19. 3 cards are drawn from a pack. The hand is known to include at least 2 aces What is the probability that it contains 3 aces? If this hand were known to include the 2 black aces, what is the probability that it contains 3 aces? [1/73, 1/25]

20. A box contains 23 ball bearings, 8 of size A, 3 of size B and 12 of size C. 3 are selected at random. What is the probability that the 3 are (*a*) of size A, (*b*) of the same size, (*c*) of different sizes? [8/253, 277/1 771, 288/1 771]

21. Show that, if 5 cards are drawn successively from a pack, the probability that there will be exactly two aces is approximately 0.040. If each card is replaced and the pack is shuffled before the next card is drawn, show that the probability of drawing exactly two aces is approximately 0.047.

22. Two persons A and B alternately cut a pack of cards, shuffling after each cut A starts and the first to cut a diamond wins. What are the probabilities of A and B winning? [4/7, 3/7]

5.17 *Bayes' Theorem*

This theorem has applications to a particular class of problems involving conditional probabilities. The following is a simple typical case for which a solution from first principles is given.

Two boxes contain respectively 3 white and 2 black balls and 2 white and 4 black balls. One of the boxes is selected, the probability of selecting the first box being 3/4. A ball is drawn from the box selected. If this ball is black, what is the probability that it came from the first box?

Let A be the event of selecting the first box and let $\bar{A}$ be the complementary event of selecting the second box. Therefore $p(A) = 3/4$ and $p(\bar{A}) = 1/4$.

Let B be the event of drawing a black ball and let $\bar{B}$ be the complementary event of drawing a white ball. Therefore $p(B/A) = 2/5$ and $p(B/\bar{A}) = 4/6 = 2/3$. It is required to calculate $p(A/B)$.

By eqn (5.50),

$$p(A/B) = p(A \cap B)/p(B) \tag{5.58}$$

But by eqn (5.51),

$$p(A \cap B) = p(A)p(B/A) = \tfrac{3}{4} \times \tfrac{2}{5} = 3/10$$

Similarly

$$p(\bar{A} \cap B) = p(\bar{A})p(B/\bar{A}) = \tfrac{1}{4} \times \tfrac{2}{3} = 1/6$$

The event B will occur only when either of the mutually exclusive events $A \cap B$ or $\bar{A} \cap B$ occurs. Therefore

$$p(B) = p(A \cap B) + p(\bar{A} \cap B) = \tfrac{3}{10} + \tfrac{1}{6} = 28/60$$

By eqn (5.58),

$$p(A/B) = \frac{3}{10}\bigg/\frac{28}{60} = \frac{9}{14}$$

The above example belongs to a class of problems in which the outcome of an experiment can result from any one of a number of independent events or "causes". It is required to calculate the probability that this outcome has resulted from the occurrence of a particular one of these events.

Problems of this type may conveniently be solved by applying the concept of sample space in which probability measures are represented by areas.

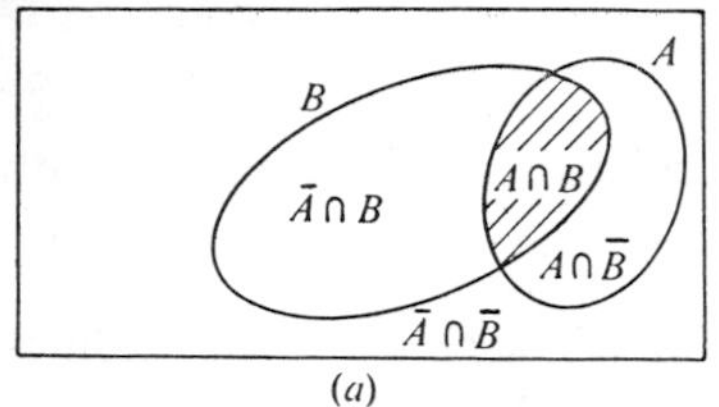

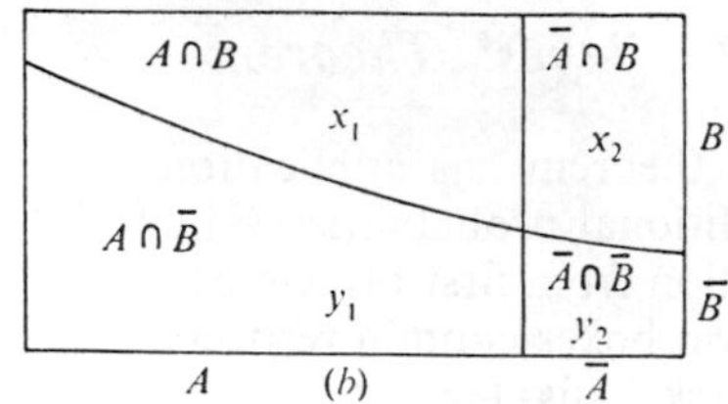

Figure 5.6

For the example above, let the contours A and B (Fig. 5.6(a)) enclose the sample spaces corresponding to the events A and B. It is convenient to represent the complete sample space (or universal probability set) by a rectangle of unit area (Fig. 5.6(b.)) The rectangles on the bases A, $\bar{A}$ have areas $\frac{3}{4}$, $\frac{1}{4}$ respectively, representing the probabilities of the selection of the first and second boxes. The curved line divides the unit rectangle into two areas representing $p(B)$ and $p(\bar{B})$ which are initially unknown.

The rectangle is thus divided into four areas representing the probability measures for $A \cap B$, $\bar{A} \cap B$, $A \cap \bar{B}$, $\bar{A} \cap \bar{B}$. The probability of the simultaneous occurrence of the events A and B will be the sum of the probabilities associated with the sample points in the region $A \cap B$. Corresponding interpretations may be given to the total probabilities assigned to the remaining regions $A \cap \bar{A}$, $\bar{B} \cap B$ and $\bar{A} \cap \bar{B}$. Let the symbols x_1, x_2, y_1 and y_2 denote these total probabilities. From Fig. 5.6(b), it follows immediately that

$$p(A) = x_1 + y_1 = 3/4 \quad \text{and} \quad p(B/A) = x_1/(x_1 + y_1) = 2/5$$
$$p(\bar{A}) = x_2 + y_2 = 1/4 \quad \text{and} \quad p(B/\bar{A}) = x_2/(x_2 + y_2) = 2/3$$

Therefore

$$x_1 = p(A)p(B/A) = \tfrac{3}{4} \times \tfrac{2}{5} = 3/10$$
$$x_2 = p(\bar{A})p(B/\bar{A}) = \tfrac{1}{4} \times \tfrac{2}{3} = 1/6$$

Required probability is

$$p(A/B) = x_1/(x_1 + x_2) = \frac{3}{10}\Big/\left(\frac{3}{10} + \frac{1}{6}\right) = \frac{9}{14}$$

An extension of the sample space method for n initial or causal events $A_1, A_2, \ldots, A_n$ leads to Bayes' Formula by which problems of this type may be solved systematically.

In Fig. 5.7 the unit rectangle is subdivided by vertical lines into the n probability spaces corresponding to the n alternative causal events A_1,

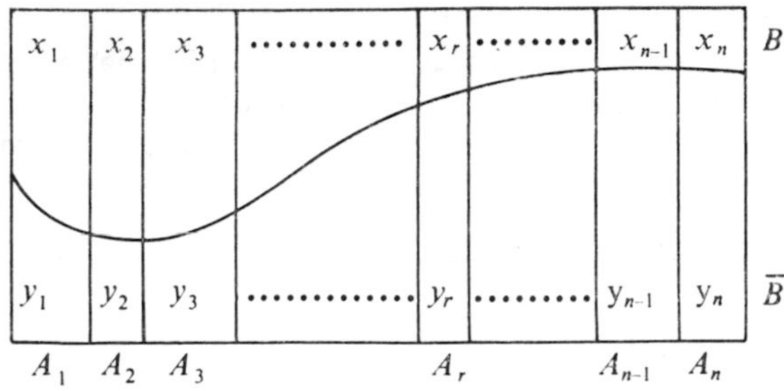

Figure 5.7

$A_2, \ldots, A_n$. As in Fig. 5.6(*b*), B and $\bar{B}$ represent the event which has occurred after the experiment has been performed and its alternative.

From Fig. 5.7, it follows that

$$\begin{array}{lll}
p(A_1) = x_1 + y_1 & p(B/A_1) = x_1/(x_1 + y_1) & \text{i.e. } x_1 = p(A_1)p(B/A_1) \\
p(A_2) = x_2 + y_2 & p(B/A_2) = x_2/(x_2 + y_2) & \text{i.e. } x_2 = p(A_2)p(B/A_2) \\
\vdots & \vdots & \vdots \\
p(A_r) = x_r + y_r & p(B/A_r) = x_r/(x_r + y_r) & \text{i.e. } x_r = p(A_r)p(B/A_r) \\
\vdots & \vdots & \vdots \\
p(A_n) = x_n + y_n & p(B/A_n) = x_n/(x_n + y_n) & \text{i.e. } x_n = p(A_n)p(B/A_n)
\end{array}$$

If the outcome is B, the probability that the event A_r has occurred is

$$p(A_r/B) = \frac{x_r}{\sum_{i=1}^{n} x_i} = \frac{p(A_r)p(B/A_r)}{\sum_{i=1}^{n} p(A_i)p(B/A_i)} \tag{5.59}$$

This is Bayes' Formula.

In some cases, the initial causal events are equally probable or it may be reasonable to assume they are. In such cases, the probabilities $p(A_i)$ are equal so that Bayes' Formula simplifies to

$$p(A_r/B) = \frac{p(B/A_r)}{\sum_{i=1}^{n} p(B/A_i)} \tag{5.60}$$

EXAMPLE 5.18

(A) A tennis tournament takes place in June. The probability that a certain player will win on a fine day is 0.8 and on a wet day is 0.6. The probability of a wet day in June is 0.2. If the player wins a game during the tournament, what was the probability that it rained on that day?

Solution Let A, $\bar{A}$ be the alternative events of a fine and a wet day.

$$p(A) = 0.8 \quad \text{and} \quad p(\bar{A}) = 0.2$$

Let B, $\bar{B}$ be the alternative events of the player winning and losing.

$$p(B/A) = 0.8 \quad \text{and} \quad p(B/\bar{A}) = 0.6$$

We wish to calculate $p(\bar{A}/B)$. Using Bayes' Formula

$$p(\bar{A}/B) = \frac{p(\bar{A})p(B/\bar{A})}{p(A)p(B/A) + p(\bar{A})p(B/\bar{A})} = \frac{(0.2)(0.6)}{(0.8)(0.8) + (0.2)(0.6)} = 3/19$$

(B) Two towns A and B are connected by four different roads R_1, R_2, R_3 and R_4. In travelling from A to B by car, the roads are selected with probabilities 0.2, 0.3, 0.4 and 0.1 respectively. The probabilities that the car can travel from A to B in one hour along these roads are respectively 0.6, 0.5, 0.7 and 0.3. If the car does the journey in one hour, what was the probability that the road R_1 was selected?

Solution Let Q be the event that the car travels from A to B in one hour.

Let R_1 be the event that the car travels along the road R_1, etc.

$$p(R_1) = 0.2,\ p(R_2) = 0.3,\ p(R_3) = 0.4,\ p(R_4) = 0.1$$

$$p(Q/R_1) = 0.6,\ p(Q/R_2) = 0.5,\ p(Q/R_3) = 0.7,\ p(Q/R_4) = 0.3$$

Required probability

$$p(R_1/Q) = \frac{p(R_1)p(Q/R_1)}{\sum_{i=1}^{4} p(R_i)p(Q/R_i)}$$

$$= \frac{(0.2)(0.6)}{(0.2)(0.6) + (0.3)(0.5) + (0.4)(0.7) + (0.1)(0.3)} = 6/29$$

(C) The probability of winning on a certain type of gambling machine is 0.15. One of four such machines is known to be out of order and the probability of winning on this machine is 0.3. One of the machines is selected at random. If the player wins, show that the probability that he selected the abnormal machine is 2/5. If the player loses, show that the probability that he selected the abnormal machine is 14/65.

Solution Since the machines are selected at random, it may be assumed that the probabilities of selecting each machine are equal. Consequently, the simplified form of Bayes' Formula (5.60) may be used.

Therefore the probability that, if the player wins, he has selected the abnormal machine is

$$0.3/0.3 + 3(0.15) = 2/5$$

and the probability that, if the player loses, he has selected the abnormal machine is

$$0.7/0.7 + 3(0.85) = 14/65$$

EXERCISES 5.7

1. Two boxes contain respectively 1 black and 2 red balls and 2 black and 3 red balls. One of the boxes is selected at random and a ball is drawn from it. What is the probability that the first box was selected if the ball drawn was (*a*) black, (*b*) red? [5/11, 10/19]
2. In a group of women, 20% are blondes and 80% are brunettes. If 50% of the blondes and 40% of the brunettes are married, what is the probability that a married women is a blonde? [5/21]
3. In a hospital, 95% of the patients who suffer from cancer and 3% of those who do not, show a positive reaction to a cancer test. 2% of the patients in the hospital have cancer. A patient, randomly selected, reacts positively. What is the probability that this patient actually has cancer? [95/242 = 0.32 approx.]
4. A bets against B in a game of cards. The probability that A has a better hand than B is 0.2 and when A has the better hand, the probability that he will raise the bet is 0.6. If, however, A has the poorer hand, the probability that he will raise is 0.1. If A raises the bet, what is the probability that he has the winning hand? [3/5]
5. There are 4 possible answers to each question on an examination paper. Good students know 80% of the answers and poor students 40%. If a good student has the correct answer to a question, what is the probability that he was guessing? Answer the same question in the case of a poor student. [1/17, 3/11]
6. Four boxes contain respectively 3 red, 2 red and 1 black, 1 red and 2 black and 3 black balls. The probabilities of selecting these boxes are respectively 0.4, 0.2,

0.3 and 0.1. A box is selected at random and 1 red ball is drawn. (*a*) What are the probabilities that the first, second, third and fourth boxes were selected? (*b*) What are the corresponding probabilities if the ball drawn was black? [(*a*) 12/19, 4/19, 3/19, 0; (*b*) 0, 2/11, 6/11, 3/11]

7. A mouse chooses at random any one of 4 mazes to escape from a box. The probabilities that the mouse will pass through these mazes in two minutes are 0.4, 0.2, 0.3 and 0.5. If the mouse escapes in two minutes, show that the probability that he chose the first maze was 2/7 and that he chose the last maze was 5/14.

8. A carton, containing a large number of electrical components, is known to have come from one of four suppliers, Brown, Jones, Green and Smith. It is known that Brown never supplies a defective component and that the other three supply 5%, 20% and 50% defectives respectively. The only method of checking a component is by destructive testing.
   A random sample of 3 components gives 1 defective and 2 non-defectives. What is the probability that the carton came from Jones? [1/15]
   Recalculate the probability given the prior information that the carton itself was selected at random from a warehouse when there were 500 cartons from Brown, 2 500 from Jones, 500 from Green and 500 from Smith [1/15, 5/19].

5.18 Tree Diagram

Consider a finite sequence of experiments such that the outcomes of each experiment, assumed finite in number, have certain probabilities. Such a sequence is called a *stochastic process* since the various outcomes depend on chance.*

Since there are several alternative outcomes for each experiment, a sequence of such experiments will result in various series of outcomes to each of which it is sought to give a probability measure so that predictions can be made for the stochastic process as a whole.

The study of such processes can be simplified by the use of schematic diagrams which show the set of all possible outcomes of the sequence of experiments.

Let us consider a sequence of three experiments, the stochastic process for which is illustrated by the tree diagram in Fig. 5.8.

Let the first experiment have two possible outcomes a, b only for which the probabilities are p_a, p_b respectively. Let c, d, e be the three possible outcomes for the second experiment assuming that a was the outcome of the first experiment, and let f, g be the two possible outcomes of the second experiment assuming that b was the outcome of the first experiment. Let h, j, k be the

* Greek word "stochos" means "guess".

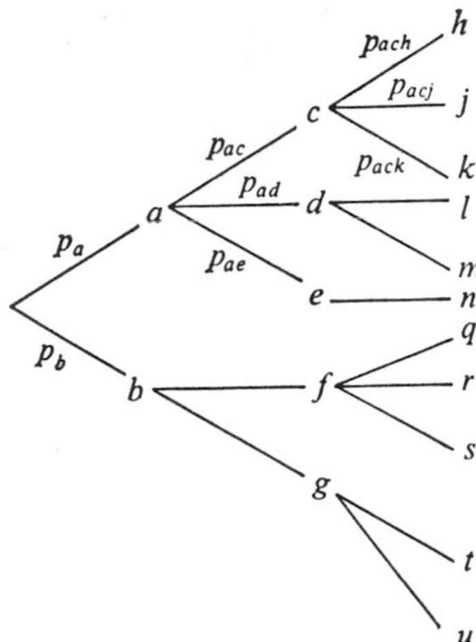

Figure 5.8 Tree Diagram

three outcomes of the third experiment assuming that a, c were the outcomes of the first two experiments, and so on.

Let p_{ac} be the probability that the outcome c occurs in the second experiment, the outcome a having occurred in the first experiment. p_{ad}, p_{ae} have similar meanings.

Let p_{ach} denote the probability of the outcome h in the third experiment, the outcomes a, c having occurred in the first and second experiments, and so on.

Since the outcomes of the first experiment are either a or b, we have

$$p_a + p_b = 1$$

Similarly, for the second and third experiments,

$$p_{ac} + p_{ad} + p_{ae} = 1 \qquad p_{ach} + p_{acj} + p_{ack} = 1 \qquad \text{and so on}$$

The probability measure of the sequence of outcomes a, c, h is $p_a p_{ac} p_{ach}$. Similarly, probability measures can be assigned to each sequence of outcomes which is represented by the various branches of the tree diagram.

The sum of the probabilities in all those branches for which the outcomes of the first two experiments were a and c is

$$p_a p_{ac} p_{ach} + p_a p_{ac} p_{acj} + p_a p_{ac} p_{ack} = p_a p_{ac}(p_{ach} + p_{acj} + p_{ack}) = p_a p_{ac}$$

This is the probability of obtaining the outcome c in the second experiment having obtained the outcome a in the first. Similar results would be obtained for any other pair of outcomes in the first two experiments.

Again the sum of the probabilities in all those branches for which the outcome of the first experiment was b is

$$p_b p_{bf} + p_b p_{bg} = p_b(p_{bf} + p_{bg}) = p_b$$

EXAMPLE 5.19

The percentages of the electorate in two towns A and B who vote Conservative, Labour and Liberal are as follows:

Town	Conservative	Labour	Liberal
A	50	40	10
B	30	50	20

One of the towns is chosen at random and two voters are chosen randomly and successively from that town. Construct a tree diagram and estimate the probability that (*a*) both voters are Labour, (*b*) the second voter chosen was Liberal.

Solution (*a*) The required probability is the probability of selecting town A and then selecting 2 Labour voters from it *plus* the probability of selecting town B and then selecting 2 Labour voters from this. From Fig. 5.9, it can

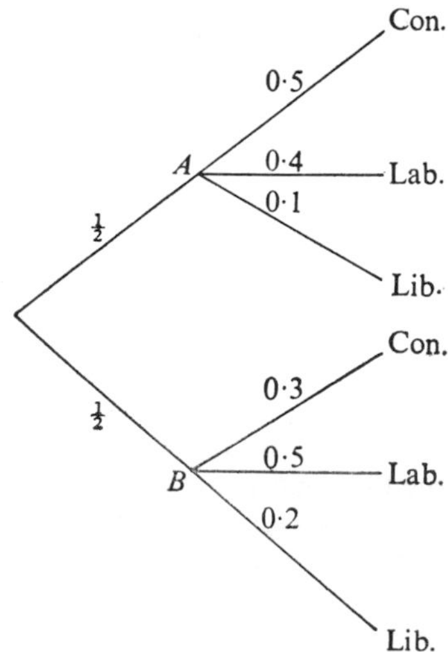

Figure 5.9

be seen that

$$\text{Required probability} = \tfrac{1}{2}(0.4)^2 + \tfrac{1}{2}(0.5)^2 = 0.205$$

(*b*) The required probability is the probability of selecting two Liberals *plus* the probability of selecting 1 Conservative then 1 Liberal *plus* the probability of selecting 1 Labour then 1 Liberal, i.e.

$$\{\tfrac{1}{2}(0.1)^2 + \tfrac{1}{2}(0.2)^2\} + \{\tfrac{1}{2}(0.5)(0.1) + \tfrac{1}{2}(0.3)(0.2)\} + \{\tfrac{1}{2}(0.4)(0.1) + \tfrac{1}{2}(0.5)(0.2)\} = 0.15$$

EXAMPLE 5.20

(A) A box contains a sample of 12 electrical components produced by a machine; 4 of the components are known to be defective; 3 components are drawn successively at random from the box, without replacement. Draw the tree diagram and find the probability that

(*a*) 2 or more defectives are drawn
(*b*) precisely 1 defective is drawn
(*c*) if the first component drawn is defective, all three are defective.

(*a*) The branches of the tree which lead to 2 or more defectives are GDD, DGD, DDG, DDD. The required probability is the sum of the probabilities in these branches, i.e.

$$\tfrac{2}{3} \times \tfrac{4}{11} \times \tfrac{3}{10} + \tfrac{1}{3} \times \tfrac{8}{11} \times \tfrac{3}{10} + \tfrac{1}{3} \times \tfrac{3}{11} \times \tfrac{8}{10} + \tfrac{1}{3} \times \tfrac{3}{11} \times \tfrac{2}{10} = 13/55$$

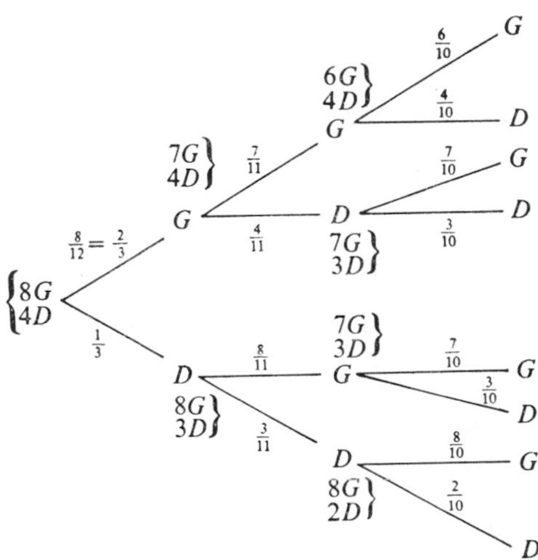

Figure 5.10 D = defective G = good

(*b*) Those branches which lead to precisely 1 defective are *GGD*, *GDG*, *DGG*. The required probability is

$$\tfrac{2}{3} \times \tfrac{7}{11} \times \tfrac{4}{10} + \tfrac{2}{3} \times \tfrac{4}{11} \times \tfrac{7}{10} + \tfrac{1}{3} \times \tfrac{8}{11} \times \tfrac{7}{10} = 28/55$$

(*c*) If the first defective has already been drawn, the probability of drawing 2 more defectives is

$$\tfrac{3}{11} \times \tfrac{2}{10} = 3/55$$

(B) Two urns X and Y contain respectively 3 black and 2 white balls and 2 black and 4 white balls. A ball is drawn at random from X and placed in Y and a ball is then drawn at random from Y. Find the probability that (*a*) both the balls drawn were of the same colour, (*b*) if the second ball drawn was white, the first ball was black.

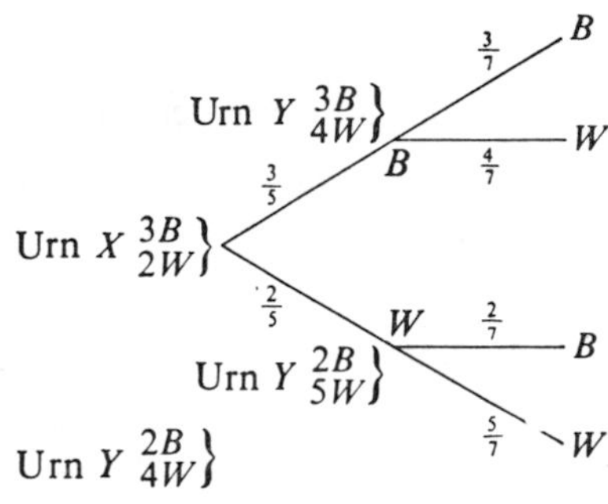

Figure 5.11 B = *black* W = *white*

(*a*) Required probability is the sum of the probabilities for *BB* and *WW*, i.e.

$$\tfrac{3}{5} \times \tfrac{3}{7} + \tfrac{2}{5} \times \tfrac{5}{7} = 19/35$$

(*b*) The probability that the event B is followed by the event W is

$$p(B \cap W) = \tfrac{3}{5} \times \tfrac{4}{7}$$

Required probability is

$$p(B/W) = \frac{p(B \cap W)}{p(W)} = \frac{\tfrac{3}{5} \times \tfrac{4}{7}}{\tfrac{3}{5} \times \tfrac{4}{7} + \tfrac{2}{5} \times \tfrac{5}{7}} = 6/11$$

5.19 Independent Trials

By adopting certain assumptions concerning the outcomes of chance experiments, the generalized treatment of stochastic processes, outlined in the last section, can be simplified to provide a mathematical analysis of the probabilities associated with experiments which occur in practice.

In this book, we shall be concerned with only two types of practical stochastic processes. In the next section, we shall consider Markov chain processes which have applications to biological and social sciences. In this section, *independent trials* processes will be considered.

Suppose that one particular experiment is repeatedly performed in such a way that the outcome of any one experiment has no effect on the outcome of any other. Let there be r outcomes $a_1, a_2, \ldots, a_r$ for each of these experiments occurring respectively with the probabilities $p_1, p_2, \ldots, p_r$ which are constant from experiment to experiment. Such a sequence of experiments comprises a process of independent trials.

In a series of independent trials, for each trial there are two outcomes only, which will be arbitrarily called "success" and "failure". The outcomes occur with constant probabilities p and q, where p is the probability of success and q that of failure and $(p + q) = 1$. The probability distribution of x successes and $(n - x)$ failures in n trials is the Binomial Distribution discussed in Section 5.12. A tree diagram for $n = 3$ is shown in Fig. 5.12.

It is required to calculate the probability of x successes and $(n - x)$ failures in n independent trials. Suppose that the tree diagram in Fig. 5.12 were extended to illustrate the case of n independent experiments. The required probability is the sum of the probabilities in all those branches which

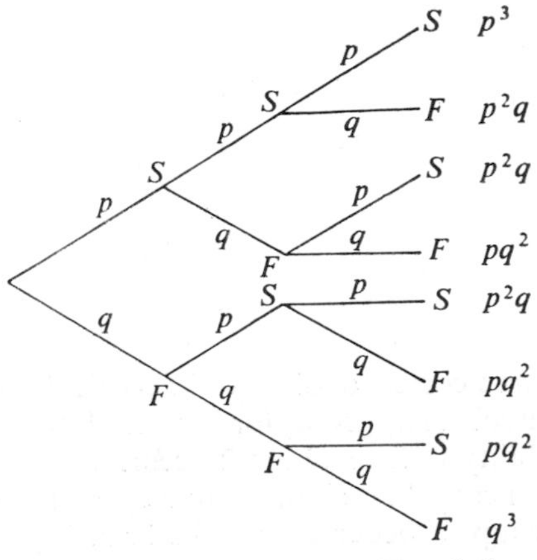

Figure 5.12 S = *success* F = *failure*

pass through x branch points marked S and $(n - x)$ marked F. The probabilities in *each* of these branches would clearly be $p^x q^{n-x}$ since every segment going towards an S has a probability p and every segment going towards an F has a probability q. It remains to find the total number of branches bearing the probability $p^x q^{n-x}$. To each such branch, there corresponds an ordered partition of the integers $1, 2, \ldots, n$ into two cells, x integers in the first and $(n - x)$ in the second. The integers in the first cell give the numbers of those trials for which S occurred and those in the second cell for which F occurred. The number of such partitions is

$$\frac{n!}{x!\,(n-x)!} \quad \text{or} \quad \binom{n}{x}$$

so that the probability of x successes in n trials is

$$\binom{n}{x} p^x q^{n-x} \qquad \text{(see eqn. 5.36)} \tag{5.61}$$

Let us now consider the general case for independent trials with more than two outcomes. For n trials, we wish to calculate the probability that there are n_1 occurrences of the outcome a_1, n_2 of $a_2, \ldots, n_r$ of a_r where $n = n_1 + n_2 + \cdots + n_r$.

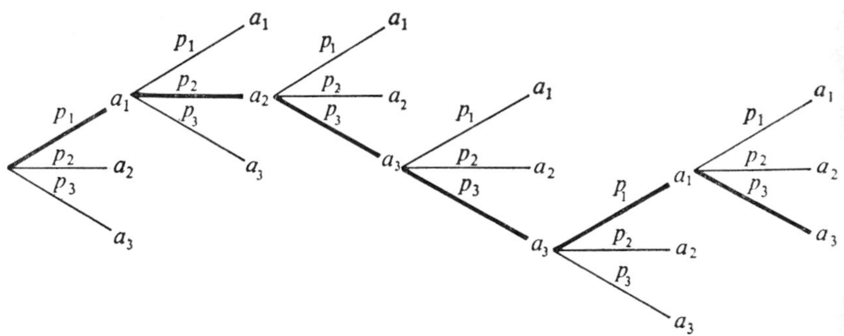

Figure 5.13

Figure 5.13 illustrates part of the tree corresponding to the particular case when $n = 6$ when there are three outcomes a_1, a_2, a_3 for each trial. Suppose that we wish to calculate the probability that there are 2 occurrences of a_1, 1 of a_2 and 3 of a_3 in 6 trials. The heavy line shows a branch of the tree diagram corresponding to such a group of occurrences. The branch probability is $p_1^2 p_2 p_3^3$.

There are other branches of the tree corresponding to 2 occurrences of a_1, 1 of a_2, and 3 of a_3, and each branch has the probability $p_1^2p_2p_3^3$. It remains to find the number of such branches. This number is the number of partitions of the integers 1, 2, . . . , 6 into three cells containing respectively 1, 2 and 3 of these integers. Thus the number of branches is 6!/(1! 2! 3!) so that

$$\text{Required probability} = \frac{6!}{1!\,2!\,3!}\,p_1^{\,2}p_2p_3^{\,3}$$

For n independent trials, it readily follows that the probability of x_1 occurrences of a_1, x_2 of a_2, . . . , x_r of a_r (where $n = x_1 + x_2 + \cdots + x_r$) is

$$f(x_1, x_2, \ldots, x_r) = \frac{n!}{x_1!\,x_2!\cdots x_r!}\,p_1^{\,x_1}p_2^{\,x_2}\cdots p_r^{\,x_r} \qquad (5.62)$$

This probability function is known as the *Multinomial Distribution* since (5.62) is the general term of the multinomial function

$$(p_1 + p_2 + \cdots + p_r)^n$$

just as the binomial probability function is the general term of the binomial function $(p + q)^n$.

As a special case of eqn (5.62), we may, for example, only be interested in finding the probability of x_1 occurrences of a_1, x_2 of a_2 irrespective of the numbers of individual occurrences of the other possible outcomes. In this case, the process may be considered to be a sequence of n independent trials having outcomes a_1, a_2, a_3', when a_3' represents any one of the outcomes $a_3, a_4, \ldots, a_r$, these three outcomes occurring respectively with probabilities p_1, p_2, p_3' where $p_3' = 1 - (p_1 + p_2)$.

$$\text{Number of occurrences of } a_3' = x_3' = n - (x_1 + x_2)$$

$$\text{Required probability} = \frac{n!}{x_1!\,x_2!\,x_3'!}\,p_1^{\,x_1}p_2^{\,x_2}p_3'^{\,x_3'} \qquad (5.63)$$

EXAMPLE 5.21
(A) A fair die is thrown 10 times. Find the probabilities that
(*a*) each face except the six comes up twice,
(*b*) the six comes up exactly 3 times and the one exactly twice.

Solution (*a*) The probability that each face comes up in any trial is 1/6. By eqn (5.62), the required probability is

$$f(2, 2, 2, 2, 2, 0) = \frac{10!}{(2!)^5}\left(\frac{1}{6}\right)^{10} = 0.0019$$

(*b*) The probability of a throw different from 1 and 6 is 4/6 and there are 5 such occurrences. By (5.63), the required probability is

$$\frac{10!}{3!\,2!\,5!}\left(\frac{1}{6}\right)^3\left(\frac{1}{6}\right)^2\left(\frac{4}{6}\right)^5 = \frac{10!}{3!\,2!\,5!}\left(\frac{1}{6}\right)^{10}4^5 = 0.043$$

(B) The probabilities that a football team will win, lose or draw, are respectively 0.7, 0.2 and 0.1. What are the probabilities that in four matches, the team will win exactly three matches and less than three matches?

Solution If the team wins exactly three matches, the result of the remaining match is either a loss or a draw. Therefore the required probability is

$$\begin{aligned} f(3, 1, 0) + f(3, 0, 1) &= \frac{4!}{3!\,1!}\{(0.7)^3(0.2)^1 + (0.7)^3(0.1)^1\} \\ &= 4(0.7)^3(0.3) = 0.412 \end{aligned}$$

For the team to win less than three matches, the required probability is

$$\begin{aligned} &f(0, 4, 0) + f(0, 3, 1) + f(0, 2, 2) + f(0, 1, 3) + f(0, 0, 4) \\ &\quad + f(1, 3, 0) + f(1, 2, 1) + f(1, 1, 2) + f(1, 0, 3) + f(2, 2, 0) \\ &\quad + f(2, 1, 1) + f(2, 0, 2) \end{aligned}$$

It is of course, simpler to calculate this probability by first calculating the total probability that the team wins exactly three and exactly four matches and subtracting from 1.

The probability of winning exactly four matches is

$$f(4, 0, 0) = (0.7)^4 = 0.2401$$

Therefore the probability of winning less than three matches is

$$1 - (0.412 + 0.240) = 0.348$$

EXERCISES 5.8

1. In a tennis tournament a man plays 3 matches alternately against two other players *A* and *B*. The probabilities that he will win against *A* and *B* are respectively 1/4 and 2/3. He plays *A* first. Find the probabilities that he will win (*a*) all 3 matches, (*b*) precisely 2 matches, (*c*) 2 consecutive matches.
   What are the corresponding probabilities if he plays *B* first?
   [1/24, 13/48, 7/24; 1/9, 4/9, 2/9]

2. The probabilities of winning on two gambling machines A and B are respectively 1/4 and 1/5. A player does not know which machine is A and selects a machine at random. Draw a tree diagram for the first game and find the probabilities that the player (*a*) wins, (*b*) chose machine B if he eventually lost. [9/40, 16/31]

3. In exercise 2, if the player plays twice, find the probability that he will win at least once if (*a*) he plays the same machine twice, (*b*) he changes machines if he loses the first time. [319/800, 2/5]

4. In a certain town, 50% of the electorate vote Labour, 35% vote Conservative and 15% vote Liberal. 6 of the voters are chosen at random. What is the probability that equal numbers support each party? [0.062]

5. Three dice are thrown simultaneously. What is the probability of a total throw of 6? [5/108]

6. A bag contains 3 red, 2 white and 5 black balls. 6 balls are drawn successively with replacement. Find the probability of drawing 2 red, 1 white and 3 black balls. [0.135]

7. A die is rolled 3 times. What are the probabilities of (*a*) a particular double, (*b*) any double? What is the probability of any three singles? [1/72, 5/12, 5/9]

8. In a game of chance, a ball is tossed into boxes which are numbered 1, 2, 3 and 4. The probabilities that the ball will fall into these boxes are 0.5, 0.25, 0.15 and 0.10 respectively. For each ball thrown into a box, the player receives as many pounds as the number on the box. Show that the probability that a player wins £5 or more in two throws is 0.2875.

9. The probability that an anti-aircraft missile will destroy an aircraft is 3/5. The probabilities of a miss and a near miss are each 1/5. Two near misses will destroy an aircraft. What is the probability that four missiles will destroy an aircraft? [124/125]

   [Hint: First find the probability that the aircraft will not be destroyed.]

5.20 Markov Chain Processes

To end this chapter, we shall deal briefly with Markov decision processes which comprise an important class of stochastic optimization problems.

Consider a sequence of stochastic experiments such that the outcome of each experiment is one of a finite number of possible outcomes $a_1, a_2, \ldots, a_r$ called *states*. The Markov process concerns a sequence such that the state of the system at any stage depends only upon its state at the immediately preceding stage. Let p_{ij} be the probability that the system is in the state a_j after an experiment given that it was in the state a_i after the preceding experiment. The symbols p_{ij} for the various transitions are called *transition probabilities* and may be conveniently exhibited in the square transition matrix

$$\mathbf{P} = \begin{pmatrix} p_{11} & p_{12} & \cdots & p_{1r} \\ p_{21} & p_{22} & \cdots & p_{2r} \\ \cdot & \cdot & & \cdot \\ \cdot & \cdot & & \cdot \\ \cdot & \cdot & & \cdot \\ p_{m1} & p_{m2} & \cdots & p_{mr} \\ \cdot & \cdot & & \cdot \\ \cdot & \cdot & & \cdot \\ \cdot & \cdot & & \cdot \\ p_{r1} & p_{r2} & \cdots & p_{rr} \end{pmatrix} \tag{5.64}$$

The sum of the transition probabilities in any row, say the mth row, of the matrix $\mathbf{P}$ is equal to 1, i.e.

$$\sum_{k=1}^{r} p_{mk} = 1$$

since, if the system is in the state a_m after any experiment, it must be in one of the states $a_1, a_2, \ldots, a_r$ after the next experiment.

DEFINITION 5.9
A stochastic matrix is a square matrix with non-negative elements such that the sum of the elements in each row is 1.

Thus $\mathbf{P}$ is a stochastic matrix. Assuming that the initial state of the system is known, $\mathbf{P}$ provides enough information to construct a tree diagram for the process from which the probabilities of various outcomes following a sequence of experiments can be calculated. As will be seen, however, Markov processes can usually be treated more conveniently by the methods of matrix algebra.

For simplicity, we shall consider three-state Markov processes only but the procedure used can be readily applied to cases of more than three states. For three states a_1, a_2 and a_3, the appropriate stochastic transition matrix is

$$\mathbf{P} = \begin{matrix} & a_1 & a_2 & a_3 \\ a_1 & p_{11} & p_{12} & p_{13} \\ a_2 & p_{21} & p_{22} & p_{23} \\ a_3 & p_{31} & p_{32} & p_{33} \end{matrix} \tag{5.65}$$

The Markov process is concerned essentially with problems of the following type. Given that the system is in the state a_i initially, what is the probability that it will be in the state a_j after n experiments? This probability will be denoted by $p_{ij}^{(n)}$ and these probabilities will be required for all possible initial and final states. For the three-state Markov chain for n experiments, these probabilities may be exhibited in the matrix form:

$$\mathbf{P}^{(n)} = \begin{matrix} & a_1 & a_2 & a_3 \\ a_1 \\ a_2 \\ a_3 \end{matrix} \!\!\! \begin{pmatrix} p_{11}^{(n)} & p_{12}^{(n)} & p_{13}^{(n)} \\ p_{21}^{(n)} & p_{22}^{(n)} & p_{23}^{(n)} \\ p_{31}^{(n)} & p_{32}^{(n)} & p_{33}^{(n)} \end{pmatrix} \tag{5.66}$$

so that if, for example, the process begins in the state a_1 the probabilities that it will be in the states a_1, a_2, a_3, after n experiments are given by the elements of the first row of $\mathbf{P}^{(n)}$.

Suppose that a three-state Markov process starts by means of a chance device which places the system in the states a_1, a_2 and a_3 with probabilities $p_1^{(0)}$, $p_2^{(0)}$ and $p_3^{(0)}$ respectively.

These initial probabilities may be represented by the vector

$$\mathbf{p}^{(0)} = (p_1^{(0)}, p_2^{(0)}, p_3^{(0)})$$

which is called a *probability vector*. Let $p_j^{(n)}$ $(j = 1, 2, 3)$ be the probability that the system will be in the state a_j after n steps. Similarly, the final probabilities may be represented by the probability vector

$$\mathbf{p}^{(n)} = (p_1^{(n)}, p_2^{(n)}, p_3^{(n)})$$

Since the probability of being in the state a_1 after n experiments equals the sum of the probabilities of being in the states a_1, a_2 and a_3 respectively after the $(n-1)$th experiment and then moving from each to the state a_1 after the nth experiment, it follows that

$$p_1^{(n)} = p_1^{(n-1)}p_{11} + p_2^{(n-1)}p_{21} + p_3^{(n-1)}p_{31}$$

Similarly

$$p_2^{(n)} = p_1^{(n-1)}p_{12} + p_2^{(n-1)}p_{22} + p_3^{(n-1)}p_{32}$$

$$p_3^{(n)} = p_1^{(n-1)}p_{13} + p_2^{(n-1)}p_{23} + p_3^{(n-1)}p_{33}$$

It follows that

$$\mathbf{p}^{(n)} = \mathbf{p}^{(n-1)}\mathbf{P} = \mathbf{p}^{(n-2)}\mathbf{P}^2 = \mathbf{p}^{(n-3)}\mathbf{P}^3 = \cdots = \mathbf{p}^{(0)}\mathbf{P}^n \tag{5.67}$$

Thus the probability vector after n experiments is obtained by multiplying the initial probability vector by the nth power of the transition matrix $\mathbf{P}$.

Now suppose that the process begins in the state a_1. In this case

$$\mathbf{p}^{(0)} = (1, 0, 0)$$

and, by eqn (5.67), the probabilities that the process will be in one of the three states a_1, a_2, a_3 after n experiments will be the elements of the first row of the matrix $\mathbf{P}^n$. Similarly, if the process begins in one of the states a_2 or a_3, the probabilities after n experiments will be the elements of the second and third rows of the matrix $\mathbf{P}^n$ respectively. With reference to eqn (5.66), it is clear that the matrix $\mathbf{P}^{(n)}$ is the nth power of $\mathbf{P}$, i.e. $\mathbf{P}^{(n)} = \mathbf{P}^n$.

EXAMPLE 5.22

(A) The transition matrix for a three-state Markov chain is

$$\mathbf{P} = \begin{matrix} & a_1 & a_2 & a_3 \\ a_1 \\ a_2 \\ a_3 \end{matrix} \begin{pmatrix} 0 & 0 & 1 \\ 2/3 & 0 & 1/3 \\ 0 & 1/4 & 3/4 \end{pmatrix}$$

Use tree diagrams for three experiments to determine the matrix $\mathbf{P}^{(3)}$. Verify by matrix multiplication that $\mathbf{P}^3 = \mathbf{P}^{(3)}$.

Solution Draw the tree diagram for three experiments (Fig. 5.14), first assuming that the process starts in the state a_1.

The probability that the process begins in the state a_1 and ends in the state a_1 is $p_{11}^{(3)} = 1 \times \frac{1}{4} \times \frac{2}{3} = 1/6$.

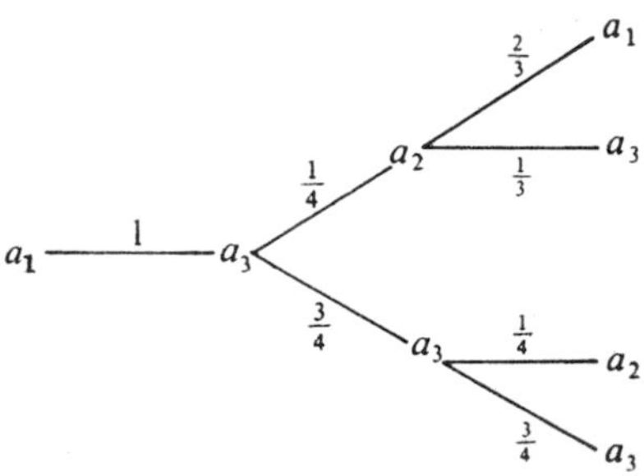

Figure 5.14

Similarly

$$p_{12}^{(3)} = 1 \times \tfrac{3}{4} \times \tfrac{1}{4} = 3/16$$

$$p_{13}^{(3)} = 1 \times \tfrac{1}{4} \times \tfrac{1}{3} + 1 \times \tfrac{3}{4} \times \tfrac{3}{4} = 31/48$$

Note that, as expected, $p_{11}^{(3)} + p_{12}^{(3)} + p_{13}^{(3)} = 1$.

These three probabilities constitute the first row of the matrix $\mathbf{P}^{(3)}$. By drawing two other tree diagrams beginning from the states a_2 and a_3, the elements of the second and third rows of the matrix $\mathbf{P}^{(3)}$ may be calculated. It will be found that

$$\mathbf{P}^{(3)} = \begin{array}{c} \\ a_1 \\ a_2 \\ a_3 \end{array} \begin{array}{c} \begin{array}{ccc} a_1 & a_2 & a_3 \end{array} \\ \begin{pmatrix} 1/16 & 3/16 & 31/48 \\ 1/18 & 11/48 & 103/144 \\ 1/8 & 31/192 & 137/192 \end{pmatrix} \end{array}$$

$$\text{Now } \mathbf{P}^2 = \begin{pmatrix} 0 & 0 & 1 \\ 2/3 & 0 & 1/3 \\ 0 & 1/4 & 3/4 \end{pmatrix} \begin{pmatrix} 0 & 0 & 1 \\ 2/3 & 0 & 1/3 \\ 0 & 1/4 & 3/4 \end{pmatrix} = \begin{pmatrix} 0 & 1/4 & 3/4 \\ 0 & 1/12 & 11/12 \\ 1/6 & 3/16 & 31/48 \end{pmatrix}$$

so that

$$\mathbf{P}^3 = \begin{pmatrix} 0 & 0 & 1 \\ 2/3 & 0 & 1/3 \\ 0 & 1/4 & 3/4 \end{pmatrix} \begin{pmatrix} 0 & 1/4 & 3/4 \\ 0 & 1/12 & 11/12 \\ 1/6 & 3/16 & 31/48 \end{pmatrix} = \begin{pmatrix} 1/16 & 3/16 & 31/48 \\ 1/18 & 11/48 & 103/144 \\ 1/8 & 31/192 & 137/192 \end{pmatrix}$$

thus verifying that $\mathbf{P}^{(3)} = \mathbf{P}^3$.

(B) The transition matrix for a three-state Markov process is

$$\mathbf{P} = \begin{pmatrix} 1 & 0 & 0 \\ 0 & 2/3 & 1/3 \\ 1/4 & 1/2 & 1/4 \end{pmatrix}$$

The initial probability vector is $\mathbf{p}^{(0)} = (1/2, 1/4, 1/4)$. Find the probabilities for the three states after two experiments.

Solution

$$\mathbf{P}^2 = \begin{pmatrix} 1 & 0 & 0 \\ 1/12 & 11/18 & 11/36 \\ 5/16 & 11/24 & 11/48 \end{pmatrix}$$

The probability vector after two experiments is

$$\mathbf{p}^{(2)} = \mathbf{p}^{(0)}\mathbf{P}^2 = (1/2 \quad 1/4 \quad 1/4)\begin{pmatrix} 1 & 0 & 0 \\ 1/12 & 11/18 & 11/36 \\ 5/16 & 11/24 & 11/48 \end{pmatrix}$$

$$= (115/192 \quad 77/288 \quad 77/576)$$

5.21 *Linear Transformation of Vectors*

From eqn (5.67) it is seen that the vector $\mathbf{p}^{(n-1)}$ is transformed into the vector $\mathbf{p}^{(n)}$ by multiplying the former by the matrix **P**. It follows that successive multiplications by **P** transform $\mathbf{p}^{(0)}$ successively into $\mathbf{p}^{(1)}$, $\mathbf{p}^{(2)}$, $\mathbf{p}^{(3)}$, ... This is an example of the linear transformation of vectors already mentioned in Chapter 9 of volume 1.

DEFINITION 5.10
A stochastic matrix is said to be *regular* if some power of the matrix has all its elements positive.

For a regular stochastic matrix a probability vector **r** can be found which transforms into itself when multiplied by **P**, i.e.

$$\mathbf{rP} = \mathbf{r}$$

If the elements of **r** are interpreted as the coordinates of a point in Euclidean space, then **r** becomes a fixed point of the transformation **P**.

The fixed point **r** may easily be found for a regular 2×2 stochastic matrix of the form

$$\mathbf{P} = \begin{pmatrix} 1 - a & a \\ b & 1 - b \end{pmatrix}$$

for suppose $\mathbf{r} = (x_1, y_1)$ is the fixed point, then

$$(x_1 \ y_1)\begin{pmatrix} 1-a & a \\ b & 1-b \end{pmatrix} = (x_1 \ y_1)$$

$$(1-a)x_1 + by_1 = x_1 \quad \text{and} \quad ax_1 + (1-b)y_1 = y_1$$

Each of these equations reduces to $ax_1 = by_1$ and since $x_1 + y_1 = 1$, we have

$$x_1 = \frac{b}{a+b} \quad \text{and} \quad y_1 = \frac{a}{a+b}$$

Thus $(b/a + b, a/a + b)$ is the unique fixed point of the transformation.

If, in the general case, the initial probability vector $\mathbf{p}^{(0)}$ is identified with $\mathbf{r}$ then

$$\mathbf{p}^{(n)} = \mathbf{p}^{(0)}\mathbf{P}^n = \mathbf{r}\mathbf{P}^n = \mathbf{r}\mathbf{P}^{n-1} = \cdots = \mathbf{r} = \mathbf{p}^{(0)}$$

so that we have a stationary Markov process in which the probabilities of the various states at any stage of the process are the same as those at the initial stage, namely $\mathbf{r}$.

The following theorems are stated, without proof.

If $\mathbf{P}$ is a regular stochastic matrix then, as $n \to \infty$,

(I) $\mathbf{P}^n \to \mathbf{R}$, a matrix of which each of the rows comprises all the elements of a unique probability vector $\mathbf{r}$.

(II) $\mathbf{p}\mathbf{P}^n \to \mathbf{r}$ where $\mathbf{p}$ is any probability vector.

The second theorem implies that no matter what the initial probabilities $\mathbf{p}^{(0)}$ of the system may be, after a large number of experiments, the probabilities of the various states will be given by the vector $\mathbf{r}$ since

$$\mathbf{p}^{(n)} = \mathbf{p}^{(0)}\mathbf{P}^n \to \mathbf{r}$$

Most problems on Markov chains may be solved by applying this important result.

EXAMPLE 5.23

Show that the matrix

$$\mathbf{P} = \begin{pmatrix} 1/2 & 1/4 & 1/4 \\ 2/5 & 3/5 & 0 \\ 1/2 & 0 & 1/2 \end{pmatrix}$$

is regular and show that its fixed point is $\mathbf{r} = (8/17, 5/17, 4/17)$. Calculate $\mathbf{P}^2, \mathbf{P}^3, \ldots$ and verify that $\mathbf{P}^n \to \mathbf{R}$.

Solution

$$\mathbf{P}^2 = \begin{pmatrix} 1/2 & 1/4 & 1/4 \\ 2/5 & 3/5 & 0 \\ 1/2 & 0 & 1/2 \end{pmatrix}\begin{pmatrix} 1/2 & 1/4 & 1/4 \\ 2/5 & 3/5 & 0 \\ 1/2 & 0 & 1/2 \end{pmatrix} = \begin{pmatrix} 19/40 & 11/40 & 1/4 \\ 11/25 & 23/50 & 1/10 \\ 1/2 & 1/8 & 3/8 \end{pmatrix}$$

All the elements of $\mathbf{P}^2$ are positive so that $\mathbf{P}$ is a regular stochastic matrix.
Let the fixed point be $\mathbf{r} = (x_1, y_1, z_1)$ so that

$$(x_1 \ y_1 \ z_1)\begin{pmatrix} 1/2 & 1/4 & 1/4 \\ 2/5 & 3/5 & 0 \\ 1/2 & 0 & 1/2 \end{pmatrix} = (x_1 \ y_1 \ z_1)$$

Therefore

$$\begin{aligned} \tfrac{1}{2}x_1 + \tfrac{2}{5}y_1 + \tfrac{1}{2}z_1 &= x_1 \\ \tfrac{1}{4}x_1 + \tfrac{3}{5}y_1 \qquad &= y_1 \\ \tfrac{1}{4}x_1 \qquad + \tfrac{1}{2}z_1 &= z_1 \end{aligned}$$

and $x_1 + y_1 + z_1 = 1$.

The solutions of these equations are $x_1 = 8/17$, $y_1 = 5/17$, $z_1 = 4/17$, so that

$$\mathbf{r} = (8/17, 5/17, 4/17) \quad \text{or} \quad (0.47, 0.29, 0.24) \text{ approx.}$$

Now

$$\mathbf{P} = \begin{pmatrix} 0.5 & 0.25 & 0.5 \\ 0.4 & 0.6 & 0 \\ 0.5 & 0 & 0.5 \end{pmatrix} \quad \text{and} \quad \mathbf{P}^2 = \begin{pmatrix} 0.48 & 0.28 & 0.25 \\ 0.44 & 0.46 & 0.10 \\ 0.50 & 0.13 & 0.37 \end{pmatrix}$$

It may be verified that

$$\mathbf{P}^3 = \begin{pmatrix} 189/400 & 227/800 & 39/160 \\ 227/500 & 193/500 & 4/25 \\ 39/80 & 1/5 & 5/16 \end{pmatrix} = \begin{pmatrix} 0.47 & 0.28 & 0.24 \\ 0.45 & 0.39 & 0.16 \\ 0.49 & 0.20 & 0.31 \end{pmatrix}$$

so that there is evidence that $\mathbf{P}^n \to \mathbf{R}$.

EXAMPLE 5.24
Taking $\mathbf{p}^{(0)} = (1/3, 1/3, 1/3)$, calculate $\mathbf{p}^{(1)}$ and $\mathbf{p}^{(2)}$ for the transition matrix $\mathbf{P}$ of the previous example and verify that they are tending towards the fixed point of $\mathbf{P}$.

Solution

$$\mathbf{p}^{(0)}\mathbf{P} = (1/3 \quad 1/3 \quad 1/3)\begin{pmatrix} 1/2 & 1/4 & 1/4 \\ 2/5 & 3/5 & 0 \\ 1/2 & 0 & 1/2 \end{pmatrix} = (7/15 \quad 17/60 \quad 1/4)$$
i.e. $(0.47 \quad 0.28 \quad 0.25)$

This is a close approximation to $\mathbf{r}$.

$$\mathbf{p}^{(0)}\mathbf{P}^2 = (1/3 \quad 1/3 \quad 1/3)\begin{pmatrix} 19/40 & 11/40 & 1/4 \\ 11/25 & 23/50 & 1/10 \\ 1/2 & 1/8 & 3/8 \end{pmatrix}$$

$$= (283/600 \quad 43/150 \quad 29/120)$$

$$= (0.47 \quad 0.29 \quad 0.24)$$

This is almost exactly the vector $\mathbf{r}$.

EXAMPLE 5.25
In a series of examinations, one of three particular questions always occurs. The same question never recurs in successive examinations. In the next examination, questions 2 and 3 are equally probable if question 1 came up in the last examination. If question 2 came up last time, question 1 is twice as probable as question 3, whilst if question 3 came up last time, question 2 is three times as probable as question 1.

Show that, in a long series of examinations, question 2 occurs most frequently and with a probability of 21/55.

Solution The transition matrix here is

$$\mathbf{P} = \begin{matrix} & Q_1 & Q_2 & Q_3 \\ Q_1 & 0 & 1/2 & 1/2 \\ Q_2 & 2/3 & 0 & 1/3 \\ Q_3 & 1/4 & 3/4 & 0 \end{matrix}$$

where the three states of the Markov chain are the choices of questions 1, 2 and 3 respectively. $\mathbf{P}$ is a regular stochastic matrix since all the elements of $\mathbf{P}^2$ are positive.

Whatever probability vector $\mathbf{p}^{(0)}$ specifies the initial probabilities of the three states Q_1, Q_2 and Q_3, after a long series of n examinations, the probability vector $\mathbf{p}^{(n)}$ specifying the final probabilities of the three states will approximate closely to the fixed point of $\mathbf{P}$.

Let the fixed point of $\mathbf{P}$ be (x_1, y_1, z_1). Therefore

$$(x_1 \ y_1 \ z_1)\begin{pmatrix} 0 & 1/2 & 1/2 \\ 2/3 & 0 & 1/3 \\ 1/4 & 3/4 & 0 \end{pmatrix} = (x_1 \ y_1 \ z_1)$$

Therefore

$$\tfrac{2}{3}x_1 + \tfrac{1}{4}z_1 = x_1$$
$$\tfrac{1}{2}x_1 + \tfrac{3}{4}z_1 = y_1$$
$$\tfrac{1}{2}x_1 + \tfrac{1}{3}y_1 = z_1$$

and

$$x_1 + y_1 + z_1 = 1$$

$$x_1 = 18/55 \qquad y_1 = 21/55 \qquad z_1 = 16/55$$

It follows that question 2 occurs most frequently.

EXAMPLE 5.26

The probability of winning with a certain type of gambling machine is 1/5. Of two such machines A and B, A is out of order and the probability of winning with it is 1/4. A player does not know which machine is A but selects one of the two machines at random. Show that if he wins his first game, the probability that he had selected A is 5/9 but if he loses his first game the probability that he had selected A is 15/31.

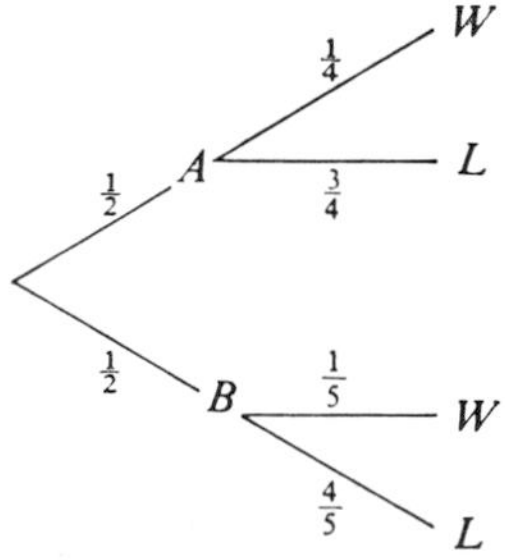

Figure 5.15 W = *win* L = *lose*

Solution If the player wins his first game, the conditional probability that he had selected A is

$$p(A/W) = p\,\frac{(A \cap W)}{p(W)} = \frac{\frac{1}{2} \times \frac{1}{4}}{\frac{1}{2} \times \frac{1}{4} + \frac{1}{2} \times \frac{1}{5}} = 5/9 > 1/2$$

If he loses his first game, the conditional probability that he had selected A is

$$p(A/L) = p\,\frac{(A \cap L)}{p(L)} = \frac{\frac{1}{2} \times \frac{3}{4}}{\frac{1}{2} \times \frac{3}{4} + \frac{1}{2} \times \frac{4}{5}} = 15/31 < 1/2$$

Since the player wishes to play the more favourable machine A, he adopts the following system in view of the values of the above conditional probabilities: if, on a particular machine, he wins his first play, he plays the same machine the next time but if he loses, he plays the other machine the next time. What is the probability that he will win in the long run?

Since, in a sequence of games, the decision, at any stage, as to how the next game shall be played depends only upon the result of the previous game, we have a two-state Markov process, the two states being the playing of machine A and of machine B. The transition matrix is

$$\begin{array}{c} \\ A \\ B \end{array}\begin{array}{c} \begin{array}{cc} A & B \end{array} \\ \begin{pmatrix} 1/4 & 3/4 \\ 4/5 & 1/5 \end{pmatrix} \end{array}$$

since, for example, if the player plays machine A, the probability that he will play A in the next game is 1/4 since this is the probability with which he wins on machine A.

The fixed vector of this matrix is easily shown to be (16/31, 15/31) so that, by adopting this system, he will play 16/31 of the time on machine A and 15/31 of the time on machine B in the long run. Thus the probability of winning in the long run is

$$\frac{16}{31} \times \frac{1}{4} + \frac{15}{31} \times \frac{1}{5} = \frac{7}{31}$$

EXERCISES 5.9

1. Show that the stochastic matrices

$$\begin{pmatrix} 1/2 & 1/2 & 0 \\ 1/3 & 1/3 & 1/3 \\ 0 & 1/4 & 3/4 \end{pmatrix} \quad \text{and} \quad \begin{pmatrix} 1/3 & 0 & 2/3 \\ 0 & 1 & 0 \\ 0 & 2/5 & 3/5 \end{pmatrix}$$

are respectively regular and non-regular.

2. A two-state Markov process is specified by the transition matrix

$$\mathbf{P} = \begin{pmatrix} 1/3 & 2/3 \\ 3/4 & 1/4 \end{pmatrix}$$

Draw tree diagrams for three experiments and hence derive the matrix $\mathbf{P}^{(3)}$. Verify that $\mathbf{P}^3 = \mathbf{P}^{(3)}$.

$$\left[\mathbf{P}^3 = \begin{pmatrix} 107/216 & 109/216 \\ 109/192 & 83/192 \end{pmatrix}\right]$$

3. Compute the first four powers of the matrix

$$\begin{pmatrix} 0.7 & 0.3 \\ 0.3 & 0.7 \end{pmatrix}$$

and hence estimate the fixed point.
[(1/2, 1/2)]

4. Show that the fixed point of all stochastic matrices of the type

$$\begin{pmatrix} 1-c & c \\ c & 1-c \end{pmatrix} \qquad (0 < c < 1)$$

is (1/2, 1/2).

5. Find the fixed point of the matrix

$$\begin{pmatrix} 1/2 & 1/2 & 0 \\ 5/16 & 1/2 & 3/16 \\ 0 & 3/4 & 1/4 \end{pmatrix}$$

[1/3, 8/15, 2/15]

6. The transition matrix of a three-state Markov process is

$$\begin{pmatrix} 0 & 1 & 0 \\ 0.3 & 0.5 & 0.2 \\ 0.4 & 0.3 & 0.3 \end{pmatrix}$$

If the initial state is specified by the probability vector (1/3, 1/6, 1/2), find the probabilities of the three states after two experiments.
[$\mathbf{p}^{(2)} = (0.243, 0.589, 0.168)$]

7. Every year 1% of the workers in the South East of England move to other jobs elsewhere whilst 3% of the workers move from other parts of England to jobs in the South East. Show that, after a long time, the proportions of all workers in the South East and elsewhere will be respectively 75% and 25% and that $\frac{3}{4}$% of all workers will be moving in each direction.

8. With reference to Example 5.26 if the player has played twice and has had two successive wins, find the probability that he had chosen machine A. What are the probabilities that he had chosen machine A having obtained the other possible pairs of outcomes win–lose, lose–win and lose–lose?
[25/41; 75/139, 75/139, 225/481]

9. With reference to Exercise 8, the player adopts a system based on the conditional probabilities obtained. He plays a machine twice. If he obtains either win–win, win–lose or lose–win, he plays two further games on the same machine. If he obtains lose–lose, he changes machines. On this system calculate the probability of winning in the long run.

$$\left[\text{Transition matrix} \begin{pmatrix} 7/16 & 9/16 \\ 16/25 & 9/25 \end{pmatrix}; \text{ fixed vector } (256/481, 225/481); \text{ probability of winning } 109/481\right]$$

6 Boolean algebras

6.1 Introduction

The publication, in 1854, of a book entitled *The Laws of Thought* by the English mathematician, George Boole (1815–1864), marked a significant advance in the development of modern algebra. He showed that, by the symbolic treatment of the premises of a given proposition, the latter could be reduced to algebraic equations from which conclusions, logically contained in the premises, could be deduced by mathematical methods, i.e. he developed a system of "symbolic logic"—a general symbolic method of logical inference. This algebra was called "Boolean Algebra" though today the abstract mathematical system designated as "Boolean Algebra" is more precisely defined. Any Boolean Algebra will contain elements, relations and operations which must obey certain definite laws which will be specified below.

6.2 The Laws of the Algebra of Sets

The algebra of sets is an example of a Boolean algebra. The laws to be obeyed by a Boolean algebra will be exemplified by those obeyed by the algebra of sets.

For any algebra of sets

(*a*) The *elements* are sets.

(*b*) The *operations* are union $\cup$, intersection $\cap$ and complementation $'$.

(*c*) The *relations* between sets or combinations of sets are equality $=$ and inclusion $\subset$.

(*a*), (*b*) and (*c*) form the structure of the algebra of sets. Now consider the properties of this structure.

1. Closure

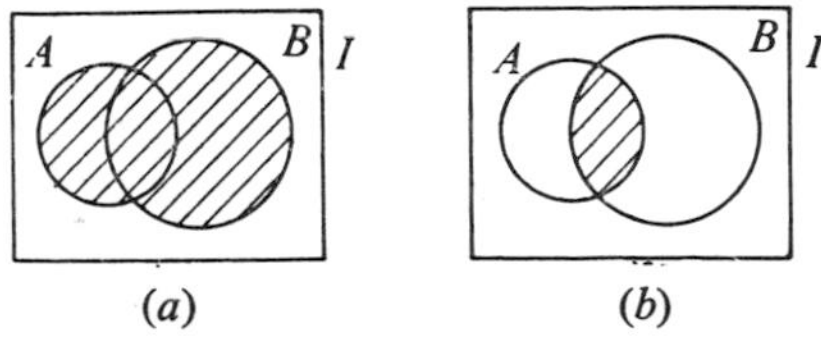

Figure 6.1 (*a*) $A \cup B$ (*b*) $A \cap B$

$$\begin{aligned} A \subset I \text{ and } B \subset I &\Rightarrow A \cup B \subset I \\ &\Rightarrow A \cap B \subset I \end{aligned}$$

i.e. if A and B are subsets of the universal set I, $A \cup B$ and $A \cap B$ are also subsets of I. The algebra of sets is said to be *closed* under union and intersection.

2. Identity Sets
(i) The identity set *relative to union* is a set whose union with any set, say A, results in the set A. The required identity set is the *empty set* $\varnothing$ since

$$A \cup \varnothing = A = \varnothing \cup A$$

(ii) The identity set *relative to intersection* is the *universal set I* since again

$$A \cap I = A = I \cap A$$
$$A \cup I = I \qquad A \cap \varnothing = \varnothing$$

3. Idempotent Laws

$$A \cup A = A \quad \text{and} \quad A \cap A = A$$

4. Complements
If A is a subset of I and A' is its complement,

$$A \cup A' = I \quad \text{and} \quad A \cap A' = \varnothing$$

Also

$$(A')' = A \qquad \varnothing' = I \qquad I' = \varnothing$$

The following laws are readily illustrated by Venn diagrams.

5. *Commutative Laws*

(i) Union $A \cup B = B \cup A$
(ii) Intersection $A \cap B = B \cap A$

6. *Associative Laws*

(i) Union $A \cup (B \cup C) = (A \cup B) \cup C$
(ii) Intersection $A \cap (B \cap C) = (A \cap B) \cap C$

7. *Distributive Laws*

(i) Union over intersection $A \cup (B \cap C) = (A \cup B) \cap (A \cup C)$
(ii) Intersection over union $A \cap (B \cup C) = (A \cap B) \cup (A \cap C)$

8. *De Morgan's Laws*
Figure 6.2 illustrates $(A \cup B)' = A' \cap B'$

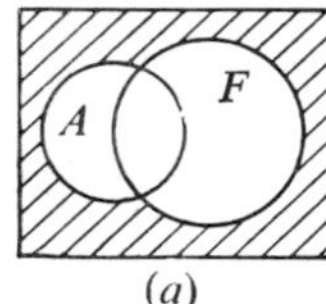

(*a*)

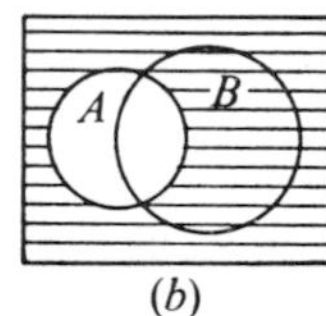

(*b*)

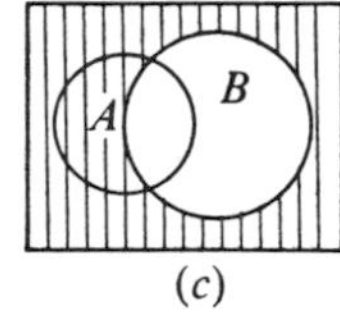

(*c*)

Figure 6.2 (*a*) $(A \cup B)'$ (*b*) A' (*c*) B'

Similarly $(A \cap B)' = A' \cup B'$. It will be observed that if $\varnothing$ and I, and $\cup$ and $\cap$, be interchanged where they occur in any one of the above laws, another law of the algebra of sets is obtained. This is the principle of duality by which a law gives rise to its dual.

6.3 *The Laws of Boolean Algebra*

The laws of the algebra of sets are a special case of the laws obeyed by the mathematical system called Boolean algebra of which the algebra of sets is a particular model. If the laws of the algebra of sets are generalized, the laws of Boolean algebra will be obtained.

Let $A, B, C, \ldots$ represent abstract elements, not necessarily sets. Let the operations $\cup$ and $\cap$ be replaced by $+$ and $\cdot$ respectively. The operations $+$ and $\cdot$ though called addition and multiplication must not be considered to possess the meanings which they have in ordinary algebra. They will represent operations which are defined for the model under consideration. Similarly

$\varnothing$ and I will be replaced by 0 and 1. The elements 0 and 1 will have only those properties which are prescribed by the laws.

If these replacements be made in the laws of the algebra of sets, then the following laws for Boolean algebra will be obtained.

1. Closure

If A and B are any elements of the Boolean algebra, $A + B$ and $A \cdot B$ will also be elements of that algebra.

2. Identity Laws

$$A + 0 = A \qquad A \cdot 0 = 0$$
$$A + 1 = 1 \qquad A \cdot 1 = A$$

3. Idempotent Laws

$$A + A = A \qquad A \cdot A = A$$

4. Complement Laws

$$A + A' = 1 \qquad A \cdot A' = 0$$
$$(A')' = A \qquad 0' = 1 \qquad 1' = 0$$

5. Commutative Laws

$$A + B = B + A \qquad A \cdot B = B \cdot A$$

6. Associative Laws

$$A + (B + C) = (A + B) + C$$
$$A \cdot (B \cdot C) = (A \cdot B) \cdot C$$

*7. Distributive Laws**

$$A + (B \cdot C) = (A + B) \cdot (A + C)$$
$$A \cdot (B + C) = (A \cdot B) + (A \cdot C)$$

8. De Morgan's Laws

$$(A + B)' = A' \cdot B'$$
$$(A \cdot B)' = A' + B'$$

**Notation* In the remainder of this chapter, the brackets will be omitted from such products as $(B \cdot C)$ so that, for example, $A + (B \cdot C)$ will be denoted by $A + B \cdot C$. The brackets will, however, be retained in such expressions as $A \cdot (B + C)$.

EXAMPLE 6.1

Consider a Boolean algebra of sets comprising the set of four elements $F = (I, A, A', \varnothing)$ where $I = (a, b, c)$ and $A = (a, b)$ so that $A' = (c)$.

Table 6.1 gives all the possible unions and intersections of the four elements and their complements.

$\cup$	I	A	A'	$\varnothing$
I	I	I	I	I
A	I	A	I	A
A'	I	I	A'	A'
$\varnothing$	I	A	A'	$\varnothing$

$\cap$	I	A	A'	$\varnothing$
I	I	A	A'	$\varnothing$
A	A	A	$\varnothing$	$\varnothing$
A'	A'	$\varnothing$	A'	$\varnothing$
$\varnothing$	$\varnothing$	$\varnothing$	$\varnothing$	$\varnothing$

Ele.	Ele.′
I	$\varnothing$
A	A'
A'	A
$\varnothing$	I

TABLE 6.1

The tables have been completed on the assumption that the identity, complement and commutative laws hold. It may now be readily verified from the tables that the property of closure holds for the four elements under the operations $\cup$, $\cap$ and $'$, and that the associative, distributive and De Morgan's laws also hold.

EXAMPLE 6.2

Let D comprise the set of the four positive integral divisors of 14, i.e. $D = (14, 7, 2, 1)$.

If x and y are any two elements of D, let $+$ and $\cdot$ be interpreted to mean that

$x + y$ is the L.C.M. of x and y and $x \cdot y$ is the H.C.F. of x and y

Let $x' = 14/x$.

Table 6.2 gives all possible combinations of the four elements under the operations $+$, $\cdot$ and $'$.

$+$	14	7	2	1
14	14	14	14	14
7	14	7	14	7
2	14	14	2	2
1	14	7	2	1

$\cdot$	14	7	2	1
14	14	7	2	1
7	7	7	1	1
2	2	1	2	1
1	1	1	1	1

x	x'
14	1
7	2
2	7
1	14

TABLE 6.2

It may readily be verified from these tables that all the laws of Boolean algebra are obeyed in this case. For example, consider the distributive law $A \cdot (B + C) = A \cdot B + A \cdot C$.

When $A = 2$, $B = 7$, $C = 14$,

$$A \cdot (B + C) = 2 \cdot (7 + 14) = 2 \cdot 14 = 2$$

$$A \cdot B + A \cdot C = 2 \cdot 7 + 2 \cdot 14 = 1 + 2 = 2$$

An examination of the tables for $F = (I, A, A', \varnothing)$ and $D = (14, 7, 2, 1)$ will show that they have the same structure. In fact, the tables for F will be transformed into the tables for D if we replace

F	I	A	A'	$\varnothing$	$\cup$	$\cap$	$'$
by							
D	14	7	2	1	$+$	$\cdot$	

There is a complete one-to-one correspondence between the Boolean algebras of F and D, which are said to be *isomorphic*.

Any statement concerning the elements of F must have a corresponding statement concerning the elements of D. For example

$(A \cap A') \cup A'$	$(7 \cdot 2) + 2$
$= \varnothing \cup A'$	$= 1 + 2$
$= A'$	$= 2$

It will be shown later that the Boolean algebras which are applicable to switching circuits and to the logic of statements are isomorphic to a simple Boolean algebra of sets. It follows that the theory of sets may be applied to circuitry and logic. This example indicates an important reason why mathematicians are interested in *mathematical structures*.

6.4 Binary Boolean Algebra

The simplest Boolean algebra contains only two elements, 0 and 1. With reference to the Laws of Boolean algebra, this means that the variables $A, B, C, \ldots$ each take the values 0 and 1 only.

This system is called *binary* Boolean algebra. A binary Boolean algebra with two variables A and B under the operations $+$, $\cdot$ and $'$ is defined by Table 6.3

$+$	$A=0$	$A=1$
$B=0$	0	1
$B=1$	1	1

$\cdot$	$A=0$	$A=1$
$B=0$	0	0
$B=1$	0	1

A	A'
0	1
1	0

TABLE 6.3

An algebra of sets consisting of two elements $\varnothing$ and I is a model of a binary Boolean algebra, and union, intersection and complementation are defined by Table 6.4.

$\cup$	$\varnothing$	I
$\varnothing$	$\varnothing$	I
I	I	I

$\cap$	$\varnothing$	I
$\varnothing$	$\varnothing$	$\varnothing$
I	$\varnothing$	I

Ele.	Ele.$'$
$\varnothing$	I
I	$\varnothing$

TABLE 6.4

The complete one-to-one correspondence between these systems, i.e. their isomorphism, is obvious. It may readily be verified that all the laws of Boolean algebra are satisfied by such binary systems, e.g. De Morgan's Laws are verified in Table 6.5.

A	B	$A+B$	$A \cdot B$	A'	B'	$A'+B'$	$A' \cdot B'$	$(A+B)'$	$(A \cdot B)'$
1	1	1	1	0	0	0	0	0	0
1	0	1	0	0	1	1	0	0	1
0	1	1	0	1	0	1	0	0	1
0	0	0	0	1	1	1	1	1	1

TABLE 6.5

From the last four columns, it will be seen that

$$(A + B)' = A' \cdot B' \qquad (A \cdot B)' = A' + B'$$

EXAMPLE 6.3

By applying the laws of Boolean algebra, show that

(i) $(A' \cdot B')' + (A' \cdot B)' = 1$
(ii) $(A + B) \cdot (A + B') = A$
(iii) $B \cdot \{(C + D)' + B\} = B$

(i) $(A' \cdot B')' + (A' \cdot B)' = A + B + A + B' = (A + A) + (B + B')$
$$= A + 1 = 1$$

(ii) $(A + B) \cdot (A + B') = (A + B) \cdot A + (A + B) \cdot B'$
$$= A \cdot A + B \cdot A + A \cdot B' + B \cdot B'$$
$$= A + B \cdot A + A \cdot B' + 0$$
$$= A + A \cdot (B + B') = A + A \cdot 1 = A + A = A$$

(iii) $B \cdot \{(C + D)' + B\} = B \cdot (C' \cdot D' + B) = B \cdot C' \cdot D' + B \cdot B$
$$= B \cdot C' \cdot D' + B$$
$$= B \cdot (C' \cdot D' + 1) = B \cdot 1$$
$$= B$$

EXERCISES 6.1

1. S is the set of integral divisors of an integer n. For any elements $x, y \in S$, $x + y$ and $x \cdot y$ are respectively defined to be the L.C.M. and the H.C.F. of x and y and $x' = n/x$.

   Determine whether S is a model of a Boolean Algebra under these operations if n is (*a*) 6, (*b*) 10, (*c*) 12.

2. If A, B and C take the values 0, 1 only, evaluate the following polynomials by tabulation as at the end of Section 6.4.

   (i) $A \cdot B' + C$ (ii) $(A + B) \cdot (A + C)$ (iii) $A \cdot (A \cdot B' + A) + A \cdot B$

   (Note that for three variables A, B, C, there will be $2^3 = 8$ possible groups of values.)

3. Use the tabular method to verify the following results for a binary Boolean algebra

   (i) $A + A \cdot B = A$

   (ii) $A + B = (A' \cdot B')'$ $\quad A \cdot B = (A' + B')'$ (De Morgan's Laws)

   (iii) $A + B \cdot C = (A + B) \cdot (A + C)$ $\quad A \cdot (B + C) = A \cdot B + A \cdot C$ (Distributive Laws)

   (iv) $(A + B) \cdot (A' + C) \cdot (B + C) = (A + B) \cdot (A' + C)$

4. By applying the laws of Boolean algebra, show that

   (i) $A \cdot (A \cdot B)' = A \cdot B'$

   (ii) $(A \cdot B' + C + A') \cdot B = B \cdot (C + A')$

   (iii) $A \cdot B + A' \cdot B' = (A' + B) \cdot (A + B')$

   (iv) $A + A' \cdot C + B = A + B + C$

   (v) $(A + B) \cdot (A' + C) + B \cdot (B' + C') = A \cdot C + B$

6.5 *Switching Circuits*

A binary Boolean algebra may be interpreted as "the algebra of circuits". In recent years, this algebra has been widely applied in the design of electronic computers and telephone dialling systems.

Suppose two electrical terminals are connected through a number of switches $A, B, C, \ldots$ in series (Fig. 6.3). When a switch is closed, current may

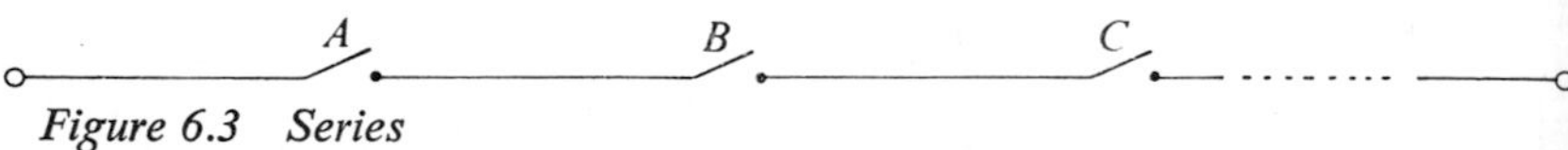

Figure 6.3 Series

pass through it but, when open, no current flows. Thus current flows between the two terminals only when *all* the switches $A, B, C, \ldots$ are closed. For switches $A, B, C, \ldots$ in series we write $A \cdot B \cdot C \ldots$, *a conjunction* of $A, B, C, \ldots$. If the terminals are connected by switches $A, B, C, \ldots$ in parallel (Fig. 6.4), the current flows if at least *one* of the switches is closed.

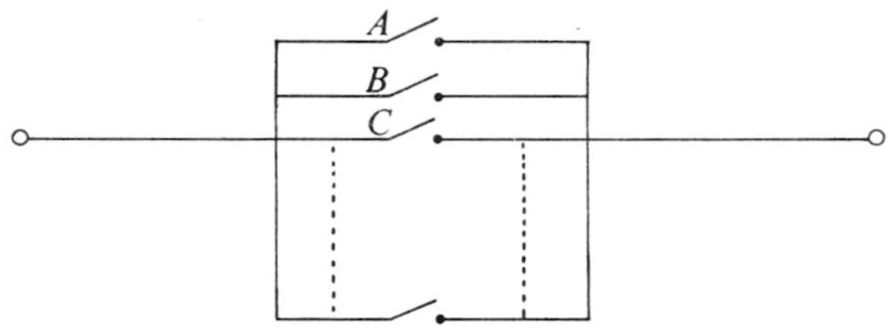

Figure 6.4 Parallel

For switches $A, B, C, \ldots$ in parallel, we write $A + B + C + \cdots$, a *disjunction* of $A, B, C, \ldots$.

Let 1 denote that a given switch is closed and let 0 denote that it is open. Consider the cases of a pair of switches A and B in series and in parallel (Fig. 6.5). For various positions of the switches A and B, all the cases which arise are listed in Table 6.6.

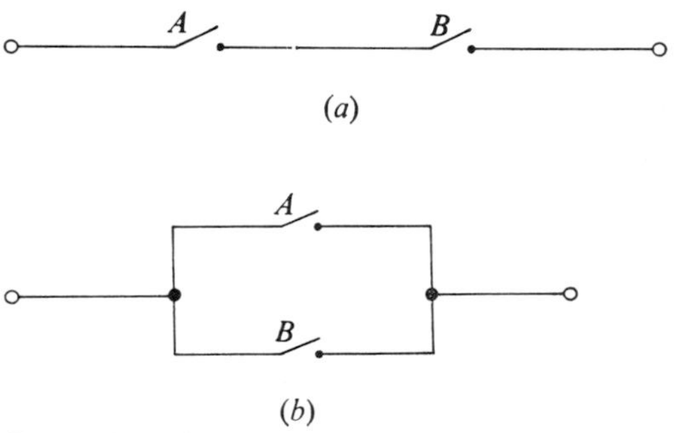

Figure 6.5 (a) *Series:* $A \cdot B$ (b) *Parallel:* $A + B$

Switch A	Switch B	Parallel $A + B$ (Disjunction)	Series $A \cdot B$ (Conjunction)
1	1	1	1
1	0	1	0
0	1	1	0
0	0	0	0

TABLE 6.6

For two switches, there are $2^2 = 4$ possible conditions. These are listed in the four lines in the first two columns of the table. The appearance of 1 in the column headed "Parallel" indicates that, for a particular arrangement of the switches A and B, the parallel circuit passes current, i.e. is closed. An entry of 0 means that the contrary is true. The appearance of 1 and 0 in the column headed "Series" is similarly interpreted.

Let A' (the complement of A) be a switch which is open when A is closed and vice versa, i.e. A' has the value 0 when A has the value 1 and vice versa.

These results may be tabulated as in Table 6.7.

		A	
	$+$	0	1
B	0	0	1
	1	1	1

Parallel

		A	
	$\cdot$	0	1
B	0	0	0
	1	0	1

Series

A	A'
0	1
1	0

TABLE 6.7

The algebra of circuits has precisely the structure of the simplest binary Boolean algebra in two variables A and B, taking the values 0 and 1 only, which was introduced above. It is likewise isomorphic with the algebra of sets, consisting of two elements $\varnothing$ and I, referred to above (Tables 6.3 and 6.4).

The laws of Boolean algebra may therefore be applied to the variables A, B, C, . . . of the algebra of circuits, for example

$$A + B \cdot C = (A + B) \cdot (A + C)$$

Interpreting this law of Boolean algebra in terms of electrical circuits, the circuits in Fig. 6.6 must be equivalent.

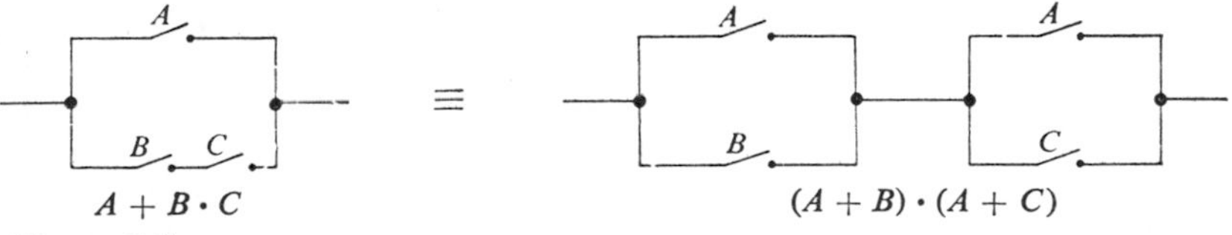

Figure 6.6

EXAMPLE 6.4

Design a circuit to put a light on a staircase on and off by two switches A and B at the top and bottom of the staircase.

Solution Mathematically speaking, the circuit must satisfy the conditions tabulated in Table 6.8

A	B	Required condition of circuit
1	1	1 (light on)
1	0	0 (light off)
0	1	0 (light off)
0	0	1 (light on)

TABLE 6.8

Consider the values of the Boolean polynomial $A \cdot B + A' \cdot B'$ for these values of A and B. It will be seen from Table 6.9 that they meet the requirements of this problem.

A	B	$A \cdot B$	A'	B'	$A' \cdot B'$	$A \cdot B + A' \cdot B'$
1	1	1	0	0	0	1
1	0	0	0	1	0	0
0	1	0	1	0	0	0
0	0	0	1	1	1	1

TABLE 6.9

It should be noted that, for all conditions of the switches A and B, the conjunction $A \cdot B$ has the value 1 only when $A = B = 1$, being otherwise 0. The conjunction $A' \cdot B'$ has the value 1 only when $A = B = 0$. It follows that $A \cdot B + A' \cdot B'$ must satisfy the conditions of the problem. $A \cdot B$ and $A' \cdot B'$ are called the basic conjunctions for $A = B = 1$ and $A = B = 0$ respectively.

We must therefore construct the circuit corresponding to $A \cdot B + A' \cdot B'$. It is shown in Fig. 6.7.

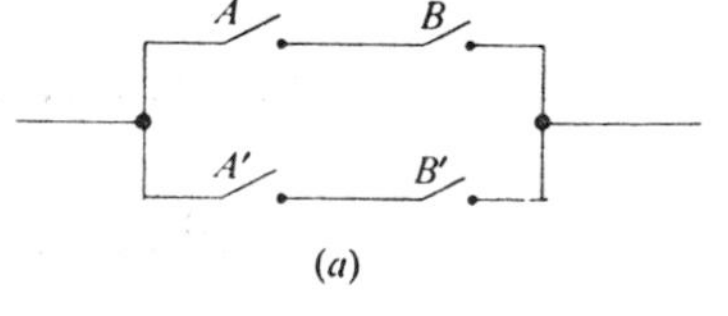

(*a*)

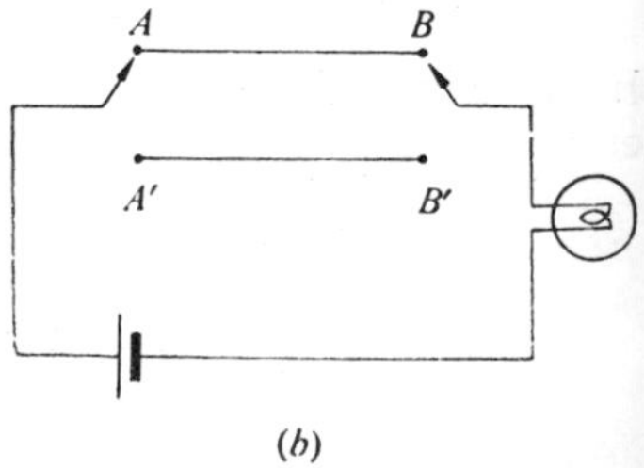

(*b*)

Figure 6.7

However since

$$A \cdot B + A' \cdot B' = (B + A') \cdot (A + B')$$

because $A' \cdot A = B \cdot B'$

it follows that an equivalent electrical circuit to that in Fig. 6.7 is the one shown in Fig. 6.8.

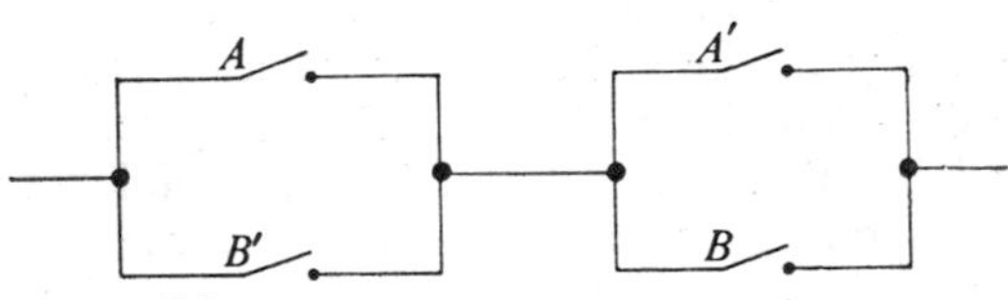

Figure 6.8

EXAMPLE 6.5
What electrical network is equivalent to the Boolean polynomial

$$(A + B) \cdot C' + A' \cdot C + B$$

By simplifying this polynomial, find the simplest equivalent circuit and verify by tabulating all possible cases.

Solution The corresponding network is shown in Fig. 6.9.

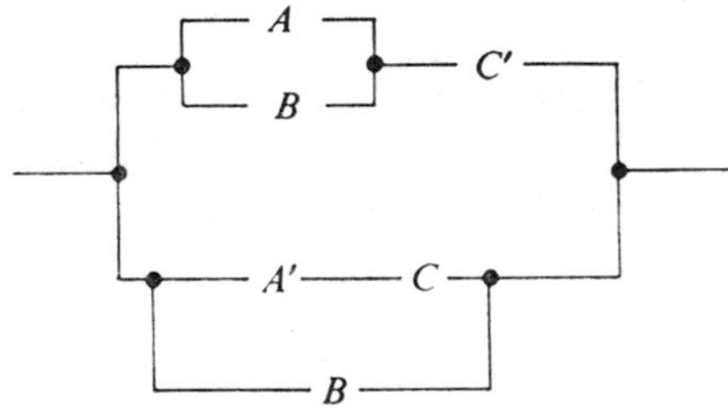

Figure 6.9

Since

$$\begin{aligned}(A + B) \cdot C' + A' \cdot C + B &= A \cdot C' + B \cdot C' + A' \cdot C + B \\ &= A \cdot C' + A' \cdot C + B(1 + C') \\ &= A \cdot C' + A' \cdot C + B\end{aligned}$$

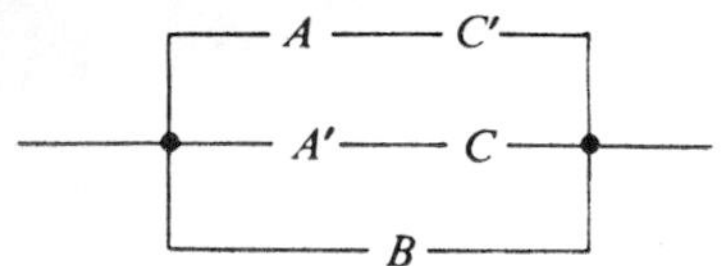

Figure 6.10

the equivalent network is as shown in Fig. 6.10. The three switches A, B, C can be set in $2^3 = 8$ possible ways. From Table 6.10 it will be seen that, for each of the eight possible settings of the switches, the values of the polynomials $(A + B) \cdot C' + A' \cdot C + B$ and $A \cdot C' + A' \cdot C + B$ agree.

This problem may be solved directly from a table giving all possible settings of the switches and the corresponding conditions which the circuit must satisfy by using an appropriate disjunction of the basic conjunctions for three switches.

If $A = B = C = 1$ (first line of Table 6.11) then the conjunction $A \cdot B \cdot C = 1$. For all other possible values of A, B, C, then $A \cdot B \cdot C = 0$.

This may be interpreted electrically as follows. If the switches A, B, C are all closed and for no other settings, the circuit represented by $A \cdot B \cdot C$, i.e. —*A*—*B*—*C*— will be closed.

Similarly the basic conjunction $A \cdot B \cdot C'$ has the value 1 when $A = B = C' = 1$ $(C = 0)$ but is 0 for all other settings of the switches. It follows that the polynomial

$$A \cdot B \cdot C + A \cdot B \cdot C' + A \cdot B' \cdot C' + A' \cdot B \cdot C + A' \cdot B \cdot C' + A' \cdot B' \cdot C$$

has the value 1 for all settings of the switches except for those in lines 3 and 8 where it has the value 0. The above polynomial therefore satisfies the conditions of this problem and represents the required circuit.

A	B	C	A'	C'	$A+B$	$(A+B)\cdot C'$	$A'\cdot C$	$A'\cdot C + B$	Values of $(A+B)\cdot C' + A'\cdot C + B$	$A\cdot C'$	Values of $A\cdot C' + A'\cdot C + B$
1	1	1	0	0	1	0	0	1	1	0	1
1	1	0	0	1	1	1	0	1	1	1	1
1	0	1	0	0	1	0	0	0	0	0	0
1	0	0	0	1	1	1	0	0	1	1	1
0	1	1	1	0	1	0	1	1	1	0	1
0	1	0	1	1	1	1	0	1	1	0	1
0	0	1	1	0	0	0	1	1	1	0	1
0	0	0	1	1	0	0	0	0	0	0	0

TABLE 6.10

A	B	C	Desired condition of the circuit	Corresponding basic conjunction
1	1	1	1	$A \cdot B \cdot C$
1	1	0	1	$A \cdot B \cdot C'$
1	0	1	0	—
1	0	0	1	$A \cdot B' \cdot C'$
0	1	1	1	$A' \cdot B \cdot C$
0	1	0	1	$A' \cdot B \cdot C'$
0	0	1	1	$A' \cdot B' \cdot C$
0	0	0	0	—

TABLE 6.11

The polynomial may now be simplified by the laws of Boolean algebra thus:

$$
\begin{aligned}
&A \cdot C' \cdot (B + B') + A' \cdot C \cdot (B + B') + B \cdot (A \cdot C + A' \cdot C') \\
&\quad = A \cdot C' + A' \cdot C + B \cdot (A \cdot C + A' \cdot C') \\
&\quad = A \cdot (C' + B \cdot C) + A' \cdot (C + B \cdot C') \\
&\quad = A \cdot (C' + B) \cdot (C' + C) + A' \cdot (C + B) \cdot (C + C') \\
&\quad = A \cdot (C' + B) + A' \cdot (C + B) \\
&\quad = A \cdot C' + A' \cdot C + B \cdot (A + A') \\
&\quad = A \cdot C' + A' \cdot C + B \quad \text{as before}
\end{aligned}
$$

EXERCISES 6.2

1. Draw the circuits represented by
   (*a*) $(A + B) \cdot (A + C)$ (*b*) $A + (B \cdot C)$
   (*c*) $(A + B') \cdot (A' + B) \cdot (A' + B')$
   (*d*) $\{A \cdot B + C'\} \cdot \{(A' + C) \cdot B\}$
   (*e*) $(A + B' + C') \cdot A' + (B + C') \cdot D$

2. What polynomials represent the circuits in Fig. 6.11? Show that these polynomials may be simplified to
   (i) $A \cdot B' \cdot C$ (ii) $A + B \cdot C$ (iii) $B \cdot C$ (iv) $A \cdot B$
   and hence draw simpler equivalent circuits. Verify by tabulation.

3. Design a circuit for a light at the foot of a staircase which is to be operated by three switches on different floors of a house.

4. Design a circuit for four switches controlling an electric lamp which is to light if (*a*) two or more switches are closed, (*b*) two or less are closed, (*c*) exactly two are closed.

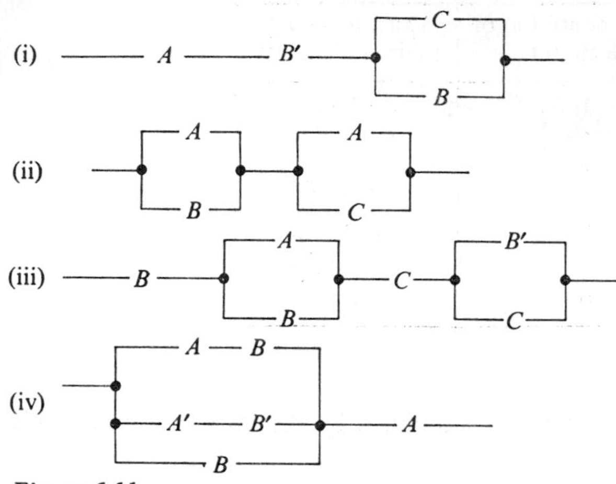

Figure 6.11

5. Each of four men in a rocket may operate his own switch controlling two lights. Design a circuit so that (*a*) a "safe" light goes on if all is ready to proceed, (*b*) a "danger" light goes on if anyone is signalling "danger".

6. A three-man Committee wishes to employ an electric circuit to indicate a secret simple majority vote. Design a circuit so that each member may push a button to indicate his "yes" vote (not push it for a "no" vote) and so that a light will go on if the majority vote "yes".
[$A \cdot B \cdot C + A \cdot B \cdot C' + A \cdot B' \cdot C + A' \cdot B \cdot C = A \cdot B + A \cdot C + C \cdot A' \cdot B$]

6.6 The Algebra of the Logic of Statements

In the algebra of sets, symbols have been used to represent sets and elements of sets. Symbols have also been used to represent electrical switches thereby formulating a Boolean algebra of "switching circuits". We shall now employ symbols in *logic* to represent *statements* or *propositions* such as "Miss X is a University student" or "Miss X is beautiful".

In this algebra, a *statement* is either *true* (T) or *false* (F). Sentences like "How do you do?" or "There are living creatures on Mars" are not statements in this context since their "truth values" are meaningless or doubtful.

Statements will be designated by $p, q, r, \ldots$ and will be used to form compound statements under the operations $\wedge$ $\vee$ and $\sim$ which mean "and", "or" and "not" respectively. These operations are analogous to $\cdot$ $+$ and $'$ of the binary Boolean algebra. The truth values T and F of statements are analogous to the values 1 and 0 taken by the variables of the binary Boolean algebra. The algebra of logic is *isomorphic* with the binary Boolean algebra.

Let p denote the statement "Miss X is a University student". Its negation, denoted by $\sim p$, would represent the statement "Miss X is not a University student". If p is true (T), $\sim p$ must be false (F), and vice versa. Let q denote the statement "Miss X is beautiful". The compound statement "Miss X is a University student and Miss X is beautiful" is denoted by $p \wedge q$ (p and q). The statement $p \wedge q$ is called the *conjunction* of p and q and is only true if *both* p and q are true and is otherwise false.

The compound statement p or q is denoted by $p \vee q$ and is called the *disjunction* of p and q. Clearly $p \vee q$ is true when *at least* one of the statements p and q is true and is false only when both p and q are false. (The use of the terms "conjunction" and "disjunction" here are directly analogous to their use in Section 6.5 for a pair of switches in series and in parallel.)

The various possibilities are summarized in the "truth tables" (Table 6.12).

p	q	$p \wedge q$
T	T	T
T	F	F
F	T	F
F	F	F

Conjunction

p	q	$p \vee q$
T	T	T
T	F	T
F	T	T
F	F	F

Disjunction

p	$\sim p$
T	F
F	T

TABLE 6.12

The isomorphism between the algebra of logic and the binary Boolean algebra is further exhibited by a comparison of Table 6.12 with the first four columns of Table 6.5.

Several compound statements of the types $p \wedge q$ and $p \vee q$ may be combined by using the basic connectives $\wedge$ $\vee$ $\sim$ and others (described below) to form more complicated compound statements, such as

$$\sim\{(p \vee q) \wedge (\sim p \vee \sim q)\}$$

The truth tables of such compound statements may be determined by a routine procedure which is illustrated for the above statement in Table 6.13.

p	q	$\sim p$	$\sim q$	$(p \vee q)$	$(\sim p \vee \sim q)$	$(p \vee q) \wedge (\sim p \vee \sim q)$	$\sim\{(p \vee q) \wedge (\sim p \wedge \sim q)\}$
T	T	F	F	T	F	F	T
T	F	F	T	T	T	T	F
F	T	T	F	T	T	T	F
F	F	T	T	F	T	F	T

TABLE 6.13

CONDITIONAL STATEMENTS

Let us now consider another connective between two statements p and q. Let p denote the statement "you work hard" and let q denote the statement "you will pass your examination".

The compound statement "if you work hard you will pass your examination" will be denoted by $p \Rightarrow q$ meaning "if p then q". The conditional statement $p \Rightarrow q$ is defined by truth Table 6.14.

p	q	$p \Rightarrow q$
T	T	T
T	F	F
F	T	T
F	F	T

TABLE 6.14

If p is the hypothesis and q the conclusion, the conditional statement $p \Rightarrow q$ is regarded as being *false* only when a false conclusion has been obtained from a *true* hypothesis, i.e. $p \Rightarrow q$ is false for the second line only. In the third line, the statement $p \Rightarrow q$ is considered to be true since no assertion has been made about what happens if "you do not work hard".

BICONDITIONAL STATEMENTS

The connective $\Leftrightarrow$ is used to denote biconditional statements. Thus $p \Leftrightarrow q$ is read "if p then q and if q then p" or "p if and only if q" and has the truth Table 6.15.

p	q	$p \Leftrightarrow q$
T	T	T
T	F	F
F	T	F
F	F	T

TABLE 6.15

From Table 6.13 it will be seen that $\sim\{(p \vee q) \wedge (\sim p \vee \sim q)\}$ has the same truth values as $p \Leftrightarrow q$ and is said to be *equivalent* to it.

Compound statements may be formed from three or more simple statements using any of the five basic connectives $\vee \; \wedge \; \sim \; \Rightarrow \; \Leftrightarrow$

$$\{p \Rightarrow (q \Rightarrow r)\} \Rightarrow \{(p \Rightarrow q) \Rightarrow (p \Rightarrow r)\}$$

is an example of a compound statement involving three simple statements p, q and r. Its truth table is worked out in Table 6.16. There are $2^3 = 8$

possible groups of truth values for the statements p, q and r so that the table has eight rows. If more than three simple statements are involved, there will be correspondingly more rows but the routine procedure for determining the truth table is unchanged.

p	q	r	$q \Rightarrow r$	$p \Rightarrow q$	$p \Rightarrow r$	$\{p \Rightarrow (q \Rightarrow r)\}$	$\{(p \Rightarrow q) \Rightarrow (p \Rightarrow r)\}$	$\{p \Rightarrow (q \Rightarrow r)\} \Rightarrow \{(p \Rightarrow q) \Rightarrow (p \Rightarrow r)\}$
T	T	T	T	T	T	T	T	T
T	T	F	F	T	F	F	F	T
T	F	T	T	F	T	T	T	T
T	F	F	T	F	F	T	T	T
F	T	T	T	T	T	T	T	T
F	T	F	F	T	T	T	T	T
F	F	T	T	T	T	T	T	T
F	F	F	T	T	T	T	T	T

TABLE 6.16

It will be observed that the compound statement is true for all possible truth values of p, q and r. Such statements are said to be *logically true*.

CONVERSE, INVERSE AND CONTRAPOSITIVE PROPOSITIONS

Having shown something of the process of the formalization of plain language into symbolic logic, let us now see how this process can help to eliminate some of the illogical reasoning which we meet so often in everyday life.

Consider the following compound statement:

"If this University is a British type of University, then this University is a good University".

Let p represent the first component and q the second. Then the above implication may be written symbolically as

$$p \Rightarrow q$$

Consider the following three variants of this implication:

$q \Rightarrow p$ "If this University is a good University, then this University is a British type of University" (Converse)

$\sim p \Rightarrow \sim q$ "If this University is not a British type of University, then this University is not a good University" (Inverse)

$\sim q \Rightarrow \sim p$ "If this University is not a good University, then this University is not a British type of University" (Contrapositive)

Let us work out their truth tables using the Table 6.16. The result is given in Table 6.17. Since $(p \Rightarrow q)$ and $(\sim q \Rightarrow \sim p)$ are both true or both false, i.e. have the same truth values irrespective of the values of p and q, it follows

p q	$\sim p$ $\sim q$	$p \Rightarrow q$	$q \Rightarrow p$	$\sim p \Rightarrow \sim q$	$\sim q \Rightarrow \sim p$
T T	F F	T	T	T	T
T F	F T	F	T	T	F
F T	T F	T	F	F	T
F F	T T	T	T	T	T

TABLE 6.17

that $(p \Rightarrow q)$ and $(\sim q \Rightarrow \sim p)$ are equivalent, i.e. a proposition is equivalent to its contrapositive. This equivalence relationship is called a *tautology*.

Similarly $(q \Rightarrow p)$ and $(\sim p \Rightarrow \sim q)$ are equivalent, i.e. the converse is equivalent to the inverse.

However, $(p \Rightarrow q)$ and $(\sim p \Rightarrow \sim q)$ are not equivalent since the two components do not have the same truth values. If, therefore, we assume that $p \Rightarrow q$ holds, it does not always follow that $\sim p \Rightarrow \sim q$. That is, for example, it is not necessarily true that "If this University is not a British type of University then this University is not a good University". In fact, it is logically incorrect to draw such a conclusion. Unfortunately, illogical reasoning of this kind is all too common.

The tautology between $(p \Rightarrow q)$ and $(\sim q \Rightarrow \sim p)$ is of considerable importance in mathematics, being widely used in the method of indirect proof. If there is difficulty in proving the proposition $(p \Rightarrow q)$ directly, it may often be proved more easily in the equivalent contrapositive form, $(\sim q \Rightarrow \sim p)$.

The algebras of sets consisting of two elements I and $\varnothing$, of circuits and of logic are all models of the binary Boolean algebra. The isomorphism between the structures of these algebras is exhibited by the examples of the one-to-one correspondences between them which are tabulated in Table 6.18.

Boolean Algebra	Sets	Circuits	Logic
$+$	$\cup$	Parallel	Disjunction
$\cdot$	$\cap$	Series	Conjunction
0	$\varnothing$	Open	F
1	I	Closed	T

TABLE 6.18

EXERCISES 6.3

1. If p = "I should like to go swimming today"
   q = "I should like to play tennis today"
   Write the following statements in words:

   (*a*) $\sim q$ (*b*) $p \vee q$ (*c*) $p \wedge q$ (*d*) $\sim(p \wedge q)$ (*e*) $\sim p \wedge q$
   (*f*) $(\sim p \wedge q) \vee (p \wedge \sim q)$

2. Write the following statements in symbolic form, letting p be "Fred is tall" and q be "George is tall".
   (*a*) Fred is tall and George is short
   (*b*) Fred and George are both short
   (*c*) Either Fred is tall or George is short
   (*d*) Neither Fred nor George is tall
   (*e*) It is not true that Fred and George are both short
   Assuming that Fred and George are both tall, which of the above compound statements are true?

3. Construct a truth table for the statement $\sim p \vee q$ and compare it with the truth table for $p \Rightarrow q$.

4. Construct truth tables for the following:

   (*a*) $\sim(p \wedge q)$ (*b*) $(p \vee \sim q) \wedge r$ (*c*) $\sim\{(\sim p \wedge \sim q) \wedge (p \vee r)\}$

   [(*a*) $F\,T\,T\,T$ (*b*) $T\,F\,T\,F\,F\,F\,T\,F$ (*c*) $T\,T\,T\,T\,T\,T\,T\,T$]

5. Show that

   (*a*) $\sim p \vee (p \vee q)$ (*b*) $p \Rightarrow (p \vee q)$ (*c*) $(p \Rightarrow q) \Leftrightarrow (\sim p \vee q)$

   are logically true but $p \wedge \sim(p \vee q)$ is logically false.

6. Show that
   (*a*) $(p \Rightarrow q) \wedge (q \Rightarrow p)$ is equivalent to $p \Leftrightarrow q$
   (*b*) $\sim p \Leftrightarrow (p \Rightarrow \sim q)$ is equivalent to $p \Rightarrow q$.
   (*c*) $\sim p \vee (\sim q \wedge \sim r)$ and $p \Rightarrow \{p \wedge \sim(q \vee r)\}$ are equivalent.

7. Verify the following tautologies by using truth tables.
   (*a*) $p \vee (q \wedge r)$ and $(p \vee q) \wedge (p \vee r)$
   (*b*) $p \wedge (q \vee r)$ and $(p \wedge q) \vee (p \wedge r)$
   (*c*) $(p \Leftrightarrow q)$ and $\{(p \Rightarrow q) \wedge (q \Rightarrow p)\}$

8. If p = "tomorrow will be a fine day", q = "we shall go to the seaside", write in words the meaning of the following statements.
   (*a*) $p \Rightarrow q$ (*b*) $q \Rightarrow p$ (*c*) $\sim p \Rightarrow \sim q$ (*d*) $\sim q \Rightarrow \sim p$
   Assuming the truth of (*a*) which of the others are true?

9. Write in words the converse, inverse and contrapositive of the following propositions:
   (*a*) If a triangle is equilateral, then it is isosceles.
   (*b*) If two lines do not intersect then they are parallel.
   (*c*) If a line joins the midpoints of two sides of a triangle then it is parallel to the third side.
   Assuming the above propositions to be true, determine the truth or falsity of their converses, inverses and contrapositives.

10. Insert $\Rightarrow$, $\Leftarrow$ or $\Leftrightarrow$ as appropriate between the pairs of statements on each line.

    $x^2 + 16 = 10x$ $(x = 2)$ or $(x = 8)$

    $x^2 + 16 = 10x$ $(x = 8)$

    $x^2 + 16 = 10x$ $(x = 2)$

11. The following "proof" depends upon incorrectly reversed implications. Find the false step.

    Let $a = b = 1$, then $a^2 = ab$. Therefore

    $a^2 - b^2 = ab - b^2$

    $(a + b)(a - b) = b(a - b)$

    $a + b = b$

    $2 = 1$

7 *Residue classes*

7.1 *Congruences*

Consider the following three subsets of the integers:

$$A = (\ldots -9, -6, -3, 0, 3, 6, 9, \ldots)$$

$$B = (\ldots -8, -5, -2, 1, 4, 7, 10, \ldots)$$

$$C = (\ldots -7, -4, -1, 2, 5, 8, 11, \ldots)$$

It will be seen that, on dividing each of the elements of subset A by 3, the remainder will be 0 in every case. When the same division is carried out for the elements of the subset B, the remainders in every case will be 1, whilst for the subset C, the remainders will all be 2.

This implies that the elements of the subsets A, B and C are respectively of the forms $3r$, $3r + 1$ and $3r + 2$ where r is an integer. It follows that the difference of any pair of elements drawn from any one of these subsets is an integer divisible by 3. For example from subset B we have

$$7 - (-8) = 15 \text{ (divisible by 3)}$$

The elements of B are said to be *congruent modulo 3*. Similarly the elements of A and C are congruent modulo 3.

DEFINITION 7.1

If x and y are two integers such that $(x - y)$ is divisible by n (a positive integer), x and y are said to be congruent modulo n. This will be expressed by the notation $x \equiv y \pmod{n}$ or simply $x \equiv y\ (n)$. [Note that $x + y\ (n)$ will be taken to mean $(x + y) \pmod{n}$.] The usual rules of arithmetic apply for the addition and multiplication of congruences. Let $a \equiv b\ (n)$ and $c \equiv d\ (n)$, and let r and s be integers. Then

$$a = b + rn \quad \text{and} \quad c = d + sn$$

$$a + c = b + d + (r + s)n$$

Therefore $a + c \equiv b + d\ (n)$ (addition)

$$ac = bd + (rd + sb)n + rsn^2$$

Therefore $ac \equiv bd\ (n)$ (multiplication)

Furthermore, for any integers λ, μ

$$\lambda a + \mu c = \lambda b + \mu d + (\lambda r + \mu s)n$$

Therefore $\lambda a + \mu c \equiv \lambda b + \mu d\ (n)$
In particular, when $\lambda = 1$, $\mu = -1$,

$$a - c \equiv b - d\ (n) \text{ (subtraction)}$$

The rule for the division of congruences is less straightforward and often involves a change of modulo. The following rule for division will now be proved.

THEOREM 7.1
If $ac \equiv bd\ (n)$ and $c \equiv d\ (n)$, then $a \equiv b\ (n/h)$ where h is the h.c.f. of d and n (or of c and n).

Note that since $c = d + sn$ (s being an integer), the h.c.f. of c and n is the same as that of d and n.

$$ac = bd + rn \quad \text{and} \quad c = d + sn$$

Therefore $ac = ad + asn = bd + rn$

$$(a - b)d = (r - as)n$$

Now $d = h\alpha$ and $n = h\beta$ where α, β are coprime (i.e. the h.c.f. of the integers α, β is 1). Therefore

$$(a - b)h\alpha = (r - as)h\beta$$

$$(a - b)\alpha = (r - as)\beta$$

Since α, β are coprime, α must be a factor of $(r - as)$.
Let $k\alpha = r - as$ where k is an integer. Therefore

$$(a - b) = k\beta = kn/h$$
$$a \equiv b\ (n/h)\ \text{(division)}$$

Corollary: If $ax \equiv bx\ (n)$, then $a \equiv b\ (n/h)$ where h is the h.c.f. of x and n.

EXAMPLE 7.1

(A) Illustrate Theorem 7.1 using $15 \equiv 190\ (7)$ and $5 \equiv 19\ (7)$.
Writing $15 \equiv 190\ (7)$ and $5 \equiv 19\ (7)$ in the form

$$ac \equiv bd\ (n) \qquad \text{and} \qquad c \equiv d\ (n)$$

gives

$$3.5 \equiv 10.19\ (7) \qquad \text{and} \qquad 5 \equiv 19\ (7)$$

The h.c.f. of 5 and 7 (i.e. h) is 1. Therefore

$$a \equiv b \left(\frac{n}{h}\right)$$

gives

$$3 \equiv 10\ (7) \qquad \text{(true)}$$

(B) Apply Theorem 7.1 to the congruence $12 \equiv 192\ (9)$ to deduce a relationship between 4 and 16. Then deduce a further relationship using the corollary.

$$12 \equiv 192\ (9) \qquad \text{i.e. } 4.3 \equiv 16.12\ (9)$$

But $3 \equiv 12\ (9)$ and the h.c.f. of 3 and 9 is 3. Therefore

$$4 \equiv 16\ (3)\ [\text{true}] \qquad \text{i.e. } 2.2 \equiv 2.8\ (3)$$

Therefore by the corollary, since h.c.f. of 3 and 2 is 1,

$$2 \equiv 8\ (3)\ [\text{true}]$$

(C) Use the congruence $12 \equiv 192\ (9)$ to illustrate the corollary to Theorem 7.1

$$1.12 \equiv 16.12\ (9)$$

The h.c.f. of 12 and 9 is 3. Therefore

$$1 \equiv 16\ (3)\ [\text{true}]$$

It is easily verified that a congruence is a reflexive, symmetric and transitive relationship between pairs of integers:

(*a*) Since $x - x = 0$, $x \equiv x\ (n)$ Reflexive

(*b*) If $x \equiv y\ (n)$, then $(x - y)$ is divisible by n. Therefore $(y - x)$ is also divisible by n, so that

$$y \equiv x(n) \qquad \text{Symmetric}$$

(*c*) If $x \equiv y\ (n)$ and $y \equiv z\ (n)$, then $x - y = rn$ and $y - z = sn$. Therefore

$$x - z = (r + s)n \qquad \text{i.e. } x \equiv z\ (n) \qquad \text{Transitive}$$

Thus $x \equiv y\ (n)$ is an *equivalence relation* (see Vol. 1, p. 79) which partitions the integers into n mutually exclusive equivalence classes each containing subsets of those integers which are *congruent modulo n*. Each of the elements of any one of these classes will give the same remainder or *residue* when divided by n. Since the possible residues are $0, 1, 2, \ldots, n - 1$, there will be n *residue classes* corresponding to modulo n. The residue class of all integers congruent to r will be denoted by **r**.

EXAMPLE 7.2

(A) The congruence relation, modulo 2, yields two residue classes

$$\mathbf{0} = (\ldots -8, -6, -4, -2, 0, 2, 4, 6, 8, \ldots)$$

$$\mathbf{1} = (\ldots -7, -5, -3, -1, 1, 3, 5, 7, 9, \ldots)$$

comprising the even and odd integers respectively.

(B) Let the integers $0, 1, 2, 3, \ldots, (n - 1)$ be uniformly spaced around a clockface. Now let n be placed at position 0, $(n + 1)$ at position 1, $(n + 2)$ at position 2, and so on. It is obvious that n residue classes, modulo n, will be found at the n positions of the clockface, and that residue class **r** will be at position r. Figure 7.1 illustrates the case for $n = 6$.

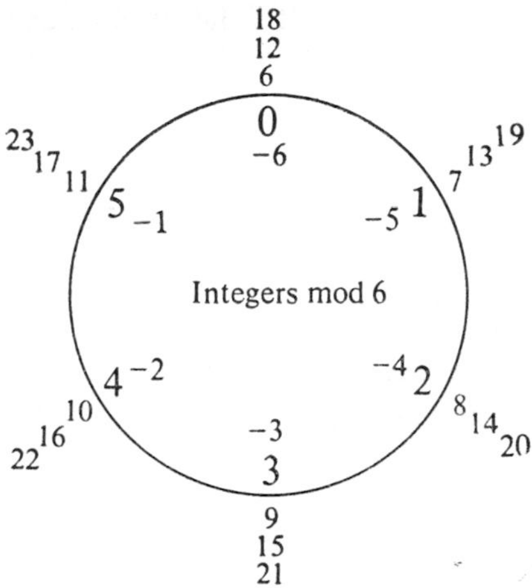

Figure 7.1

7.2 Algebra of Residue Classes

If any element of the class $\mathbf{r}$ is added to any element of the class $\mathbf{s}$, the sum will be an element of the class which contains $r + s$, since

$$(r + \lambda n) + (s + \mu n) = (r + s) + (\lambda + \mu)n$$

Since $r + s$ may exceed n, the class which contains $r + s$ will be the class $\mathbf{t}$ where $t \equiv r + s\ (n)$ and $0 \leqslant t \leqslant n - 1$ and we define $\mathbf{r} + \mathbf{s}$ to be $\mathbf{t}$, a unique residue class.

It is obvious that

$$\mathbf{r} + \mathbf{s} = \mathbf{s} + \mathbf{r}$$

$$\mathbf{r} + (\mathbf{s} + \mathbf{t}) = (\mathbf{r} + \mathbf{s}) + \mathbf{t}$$

$$\mathbf{r} + \mathbf{0} = \mathbf{r}$$

so that for the algebra of residues, having a finite number (n) of elements, the commutative and associative laws hold and a zero exists. In addition, the residue class $\mathbf{0}$ is a *neutral* element.

We define $\mathbf{r} - \mathbf{s}$ to be the class containing $r - s$.

If any element of the class $\mathbf{r}$ is multiplied by any element of class $\mathbf{s}$, the product will be an element of the class which contains rs, since

$$(r + \lambda n)(s + \mu n) = rs + \gamma n$$

We define $\mathbf{r.s}$ to be the class which contains rs. It is fairly obvious from definitions of the sum and product of residue classes, that the distributive law holds:

$$\mathbf{r.(s + t) = r.s + r.t}$$

Also since $\mathbf{1.r = r}$, the residue class $\mathbf{1}$ is a *neutral* element. The residue classes, modulo n, thus comprise a set of n elements which may be added, subtracted and multiplied like the integers $0, 1, 2, \ldots, (n - 1)$.

For integers, the equation $rx = ry$ necessarily implies either that $r = 0$ or division of each side of the equation by r, i.e. cancelling r, gives $x = y$.

This cancellation law does not necessarily hold for residues. For example (i) Modulo 11, we have

$$\mathbf{4.2 = 3.10 = 8}$$

On cancelling by 2, we have

$$\mathbf{2.2 = 3.5 = 4}$$

which is true
(ii) Modulo 12, we have

$$\mathbf{3.4 = 6.8 = 0}$$

but

$$\mathbf{3.1 \neq 6.2} \quad \text{and} \quad \mathbf{1.1 \neq 2.2}$$

so that cancellation by 4 or 12 would not be valid.

The division of one residue by another, which the cancellation process involves, is not always possible. The conditions under which division may be performed and under which cancellation is valid will be considered in the next section.

7.3 Division of Residues

It is first necessary to define the meaning of $\mathbf{s/r}$. To interpret this as the value of the quotient s/r would be unsatisfactory since s/r is not always an integer. A useful definition may be expressed in terms of the multiplication of residues.

DEFINITION 7.2
$\mathbf{s/r}$ is the value of $\mathbf{x}$ which satisfies the equation

$$\mathbf{r.x = s} \tag{7.1}$$

provided the solution exists and is unique.

Cases arise where the division $\mathbf{s}/\mathbf{r}$ is not possible either because no value of $\mathbf{x}$ exists satisfying (7.1) or because more than one such value exists. For example, modulo 6,

(i) $\mathbf{5.x} = \mathbf{4}$ has a unique solution $\mathbf{x} = \mathbf{2}$.
(ii) $\mathbf{3.x} = \mathbf{3}$ has three solutions $\mathbf{x} = \mathbf{1, 3, 5}$.
(iii) $\mathbf{4.x} = \mathbf{3}$ has no solution.

Division is therefore defined to be possible in (i) only and $\mathbf{4}/\mathbf{5} = \mathbf{2}$. The following important theorem, concerning the division of residues, will now be proved.

THEOREM 7.2
For the residue classes modulo n, division by $\mathbf{r}$ is possible if and only if r, n are coprime and $r \neq 0$.

Consider separately the cases when r and n are and are not coprime.

(1) *r, n coprime*
By Euclid's Algorithm

$$ri + nj = 1$$

where i, j are integers (see Appendix, eqn. A.2, p. 223).

For *any* integral value of s

$$ris + njs = s \qquad ris \equiv s\ (n) \qquad \mathbf{r.is} = \mathbf{s}$$

so that $\mathbf{is}$ is a solution of $\mathbf{r.x} = \mathbf{s}$ for any value of $\mathbf{s}$. It follows that if $\mathbf{x}$ takes each of the n possible values $\mathbf{0, 1, 2, \ldots, n-1}$, then $\mathbf{r.x}$ must take each of the same n different values at least once and therefore once only.

Thus $\mathbf{r.x} = \mathbf{s}$ has the unique solution $\mathbf{is}$, as, for example, in case (i) above. It follows that if $\mathbf{r.x} = \mathbf{r.y}$, then $\mathbf{x} = \mathbf{y}$ provided $\mathbf{r} \neq 0$, so that the cancellation law is valid.

(2) *r, n not coprime*
Let h be the h.c.f. of r, n and let $r = h\alpha$, $n = h\beta$.

Then $rx + kn$ is a multiple of h for *any* integral value of x.

Therefore $\mathbf{r.x} = \mathbf{t}$ where t is a multiple of h.

It follows that $\mathbf{r.x} = \mathbf{s}$ has *no* solution unless s is a multiple of h (as in case (iii) above).

If, however, s is a multiple of h, a number of solutions may be found (case (ii) above):

Let $s = h\gamma$. Then by Euclid's Algorithm

$$ri + nj = h \qquad \text{(see eqn. A.1, p. 223)}$$

$$ri\gamma + nj\gamma = h\gamma = s \qquad ri\gamma \equiv s\ (n)$$

Therefore $\mathbf{r.i\gamma} = \mathbf{s}$ so that $\mathbf{i\gamma}$ is one solution of $\mathbf{r.x} = \mathbf{s}$.

Now $r\beta = n\alpha$, so $r\beta \equiv 0\ (n)$. Therefore $\mathbf{r.\beta} = \mathbf{0}$. Therefore remembering $h\beta = n$,

$$\mathbf{r.(i\gamma} + k\boldsymbol{\beta}) = \mathbf{s} \qquad [k = 0, 1, 2, \ldots, (h-1)]$$

Thus $\mathbf{i\gamma} + k\boldsymbol{\beta}$, where $k = 0, 1, 2, \ldots, (h-1)$ and where $\beta = n/h$, are h *different* solutions of $\mathbf{r.x} = \mathbf{s}$ for any given s.

Therefore by definition, division is not possible when r, n are not coprime and, in this case, the cancellation law is not valid since $\mathbf{r.x} = \mathbf{r.y}$ does not imply that $x \equiv y$ (modulo n) though it does imply that $x \equiv y$ (modulo $\beta = n/h$) as also follows from Theorem 7.1.

It follows from (1) above that where $n = p$ and p is prime, if x takes the values $1, 2, \ldots, (p-1)$, then ax takes the same set of values in some order, provided a is not a multiple of p. Therefore

$$a.2a.3a.\ldots(p-1)a \equiv 1.2.3.\cdots(p-1) \text{ (modulo } p)$$

$$a^{p-1} \equiv 1\ (p) \quad \text{provided } a \not\equiv 0\ (p)$$

$$a^{p} \equiv a\ (p) \quad \text{provided } a \not\equiv 0\ (p)$$

But $a^p \equiv a\ (p) \equiv 0\ (p)$ even if $a \equiv 0\ (p)$. We have therefore proved that if p is prime, then $a^p \equiv a\ (p)$, and that if a is not a multiple of p, $a^{p-1} \equiv 1\ (p)$. This is *Fermat's Theorem.*

7.4 Arithmetic Prime Modulo

In this case, all the residues* $1, 2, 3, \ldots, (p-1)$ are coprime with p; therefore division by each of these residues is possible. It follows that, in this arithmetic, the rules of addition, subtraction, multiplication and division are applicable to this set of p elements and that the set is closed under these operations. Cancellation is always valid in this arithmetic.

For $p = 7$, the addition and multiplication tables are shown in Table 7.1.

In the multiplication table, it will be seen that every row and column (except the first) contains each residue once and only once. This was to be expected since $rx = s$ has a unique solution for all p values of s for a given

* Residues will be printed in ordinary italic type from here onwards, not bold.

Addition

+	0	1	2	3	4	5	6
0	0	1	2	3	4	5	6
1	1	2	3	4	5	6	0
2	2	3	4	5	6	0	1
3	3	4	5	6	0	1	2
4	4	5	6	0	1	2	3
5	5	6	0	1	2	3	4
6	6	0	1	2	3	4	5

Multiplication

×	0	1	2	3	4	5	6
0	0	0	0	0	0	0	0
1	0	1	2	3	4	5	6
2	0	2	4	6	1	3	5
3	0	3	6	2	5	1	4
4	0	4	1	5	2	6	3
5	0	5	3	1	6	4	2
6	0	6	5	4	3	2	1

TABLE 7.1

value of r since p is prime. Inverses are indicated by the positions of 1 in the table. They are seen to be 1 and 1, 2 and 4, 3 and 5 and 6 and 6.

7.5 Arithmetic Nonprime Modulo

In this case, n and one or more of the numbers 1, 2, 3, . . . , $(n - 1)$ will *not* be coprime. Consequently, whilst the rules of addition, subtraction and multiplication apply to all n elements, division is only possible by those residues which are coprime with n.

For the composite modulo $n = 6$, the addition and multiplication tables are shown in Table 7.2

Since only 1 and 5 are coprime with 6, only the rows and columns corresponding to 1 and 5 in the multiplication table will contain each of the residues 0, 1, 2, . . . , 5. Since 2, 3, 4 are not coprime with 6, division by these residues is not possible.

The residue equation $2x = 4$ has two solutions $x = 2, 5$ which is consistent with the fact that 2 is the h.c.f. of 2 and 6 and 4 is a multiple of 2.

Addition

+	0	1	2	3	4	5
0	0	1	2	3	4	5
1	1	2	3	4	5	0
2	2	3	4	5	0	1
3	3	4	5	0	1	2
4	4	5	0	1	2	3
5	5	0	1	2	3	4

Multiplication

×	0	1	2	3	4	5
0	0	0	0	0	0	0
1	0	1	2	3	4	5
2	0	2	4	0	2	4
3	0	3	0	3	0	3
4	0	4	2	0	4	2
5	0	5	4	3	2	1

TABLE 7.2

Similarly $2x = 2$ has two solutions $x = 1, 4$ whilst $2x = 1, 3, 5$ have no solutions since 1, 3, 5 are not multiples of 2.

The residue equation $3x = 3$ has three solutions $x = 1, 3, 5$ which is consistent with 3 being the h.c.f. of 3 and 6 and also a factor 3.

On the other hand, $3x = 1, 2, 4$, or 5 have no solutions since 3 is coprime with 1, 2, 4 and 5. Since 2 is the h.c.f. of 4 and 6, and 2 is a factor of 2 and 4, then $4x = 2$ has two solutions $x = 2, 5$, whilst $4x = 4$ has two solutions $x = 1, 4$, differing by $\beta = n/h = 3$. Also $4x = 1, 3, 5$ have no solutions.

EXAMPLE 7.3

(A) Solve $17x \equiv 23$ (29).

Solution Since 29 is prime, there is a unique solution. The solution may be obtained by applying Euclid's Algorithm thus:

$29 = 1 . 17 + 12$		$12x \equiv -17x \equiv -23 \equiv 6$
$17 = 1 . 12 + 5$		$5x \equiv 17x - 12x \equiv -6 - 6 \equiv -12$
$12 = 2 . 5 + 2$	so that	$2x \equiv 12x - 2 . 5x \equiv 6 + 24 \equiv 30 \equiv 1$
$5 = 2 . 2 + 1$	(mod 29)	$x \equiv 5x - 2 . 2x \equiv -12 - 2$
$2 = 2 . 1 + 0$		$\equiv -14 \equiv 15$

Solution $x \equiv 15$

(B) Solve $15x \equiv 6$ (27).

Solution Here 15, 27 are not coprime. Their h.c.f. is 3 so that there will be 3 solutions differing by $n/h = 27/3 = 9$.

$-27 = 15 + 12$		$12x \equiv -15x \equiv -6$
$15 = 12 + 3$	so that	$3x \equiv 15x - 12x \equiv 6 + 6 \equiv 12$
$12 = 4 . 3 + 0$	(mod 27)	$x \equiv 4$

The three solutions are $x \equiv 4, 13, 22$.

(C) Solve $5x^2 - 2x \equiv 11$ (13).

Solution Such equations may be solved by modifying the constant term and the term in x until each term is divisible by the coefficient of x^2, i.e. 5.

$$5x^2 - 2x + 4 . 13x - 11 + 2 . 13 \equiv 0 \ (13)$$

$$5x^2 + 50x + 15 \equiv 0 \ (13)$$

Since 5, 13 are coprime,

$$x^2 + 10x + 3 \equiv 0\ (13)$$

$$(x + 5)^2 \equiv 22 \equiv 9\ (13)$$

$$x + 5 \equiv 3 \text{ or } 10 \qquad x \equiv 11 \text{ or } 5$$

(D) Show that an integer is divisible by 11 if the sum of its digits, taken with alternate signs, is divisible by 11.

Solution

$$10 \equiv -1\ (11) \qquad 10^2 \equiv 1\ (11)$$

$$10^3 \equiv -1\ (11) \qquad 10^4 \equiv 1\ (11) \quad \text{and so on}$$

Any integer may be expressed in decimal form thus

$$n = a_0 + 10a_1 + 10^2a_2 + 10^3a_3 + \cdots + 10^na_n$$

Let $s = a_0 - a_1 + a_2 - a_3 + \cdots + (-1)^na_n$. Then

$$n - s = 11a_1 + (10^2 - 1)a_2 + (10^3 + 1)a_3 + (10^4 - 1)a_4 + \cdots$$

Now $(10^2 - 1)$, $(10^3 + 1)$, $(10^4 - 1)\ldots$ are all congruent to 0 (mod 11). Therefore

$$n - s \equiv 0\ (11) \quad \text{and the rule follows}$$

(E) The smallest positive integer c for which $a^c \equiv 1\ (p)$, where p is prime, must be a divisor of $(p - 1)$. Illustrate by taking $a = 4$, $p = 23$.

Solution Divide $(p - 1)$ by c so that $p - 1 = kc + r$ where $0 \leqslant r < c$.

Then $a^{p-1} = a^{kc} \cdot a^r \equiv 1\ (p)$ by Fermat's Theorem
Since $a^{kc} \equiv 1(p)$

$$a^r \equiv 1\ (p) \qquad r = 0 \qquad \text{so that } c \text{ is a divisor of } (p - 1)$$

$$4^3 = 64 \equiv -5\ (23) \qquad 4^4 = 256 \equiv 3\ (23) \qquad 4^8 \equiv 9\ (23)$$

$$4^{11} \equiv -45 \equiv 1\ (23) \qquad 4^{22} \equiv 1\ (23) \quad \text{and 11 is a factor of } (23 - 1).$$

EXERCISES 7.1

1. Write down the residue classes (i) mod 5, (ii) mod 4.

2. Write out the addition tables for residue classes (i) mod 4, (ii) mod 5. What are the neutral elements and does every element have an inverse in (i) and (ii)? [0, yes]

3. Write out the multiplication table for residue classes mod 5. Is there a neutral element?

   Omitting the row and column of zeros, what is the neutral element in the table? What are the inverses of 1, 2, 3, 4?

   What are the solutions of the following residue equations, mod 5?

   (i) $4x = 3$ (ii) $x^2 = 1, 4$ (iii) $x^2 = 2, 3$

   [1; 1, 3, 2, 4; (i) 2 (ii) (1, 4), (2, 3) (iii) no solutions]

4. Write out the multiplication tables for residue classes (i) mod 4, (ii) mod 8. What are the essential differences between these tables and the corresponding table for mod 5? Account for them.

5. What are the solutions of the following residue equations, mod 4?

   (i) $3x = 2$ (ii) $2x = 2$ (iii) $2x = 3$ (iv) $x^2 = 1$

   [(i) 2, (ii) 1, 3, (iii) no solution, (iv) 1, 3]

6. Solve the following congruence equations

   (i) $3x \equiv 11$ (16) (ii) $11x \equiv 2$ (19) (iii) $6x \equiv 13$ (30) (iv) $10x \equiv 22$ (31)

   (v) $4x \equiv 4$ (8) (vi) $24x \equiv 12$ (44) (vii) $6x \equiv 12$ (30) (viii) $14x \equiv 7$ (21)

   [(i) 9, (ii) 14, (iii) no solution, (iv) 27, (v) 1, 3, 5, 7, (vi) 6, 17, 28, 39, (vii) 2, 7, 12, 17, 22, 27, (viii) 2, 5, 8, 11, 14, 17, 20]

7. Solve the congruence equations

   (i) $3x^2 + 4x \equiv 1$ (17) (ii) $6x^2 - 5x \equiv 27$ (37)

   [(i) 4, (ii) 12, 32]

8. Show that an integer is divisible by 3 or 9 if the sum of its digits is divisible by 3 or 9 respectively.

9. Show that an integer is divisible by 7 if the expression

   $$s = a_0 + 3a_1 + 2a_2 - a_3 - 3a_4 - 2a_5 + \cdots$$

   is divisible by 7 where $a_0, a_1, a_2, \ldots$ are the successive digits of the integer expressed in decimal form.

10. Check Fermat's Theorem for $p = 5, 7, 11$ and 17, taking $a = 3, 4, 5, 8$.

Appendix: Euclid's Algorithm

To find the h.c.f. of two integers a, b ($a > b > 0$). Let q_1 be the quotient and r_1 the remainder when a is divided by b, then

$$a = bq_1 + r_1 \qquad b > r_1 \geqslant 0$$

Similarly, on dividing b by r_1, we have

$$b = r_1 q_2 + r_2 \qquad b > r_1 > r_2 \geqslant 0$$

Now divide r_1 by r_2 and continue to divide successive pairs of remainders. After a finite number of such divisions, say $(n + 1)$, the remainder r_{n+1} will be zero and $b > r_1 > r_2 > \cdots > r_{n+1} = 0$.

We then have the following list of equations

$$\begin{aligned}
a &= bq_1 + r_1 \\
b &= r_1 q_2 + r_2 \\
r_1 &= r_2 q_3 + r_3 \\
&\cdot \\
&\cdot \\
&\cdot \\
r_{n-3} &= r_{n-2} q_{n-1} + r_{n-1} \\
r_{n-2} &= r_{n-1} q_n + r_n \\
r_{n-1} &= r_n q_{n+1} + 0
\end{aligned}$$

From the last equation, r_n is a factor of r_{n-1}. From the preceding equation, it follows that r_n is a factor of r_{n-2}. Using these equations successively from the bottom upwards, r_n is seen to be a factor of $r_{n-3}, r_{n-4}, \ldots, r_2, r_1$ and finally of b and a.

Furthermore, if k is *any* common factor of a and b, it follows from the first equation that k is a factor of r_1 and therefore, from the second equation, k is a factor of r_2. Working downwards through the list of equations, k is a factor of r_n. Since r_n is the largest factor of r_n, r_n must be the h.c.f. of a and b.

It will now be shown that if h is the h.c.f. of a and b, there are integers λ, μ such that

$$\begin{aligned}
h &= \lambda a + \mu b \\
r_n = h &= r_{n-2} - r_{n-1} q_n \\
&= r_{n-2} - q_n(r_{n-3} - r_{n-2} q_{n-1}) \\
&= \lambda_1 r_{n-3} + \mu_1 r_{n-2} \qquad \text{where } \lambda_1, \mu_1 \text{ are integers} \\
&= \lambda_1 r_{n-3} + \mu_1(r_{n-4} - r_{n-3} q_{n-2}) \\
&= \lambda_2 r_{n-4} + \mu_2 r_{n-3} \quad \text{and so on}
\end{aligned}$$

Continuing this process, we have ultimately

$$h = \lambda a + \mu b \qquad \text{where } \lambda, \mu \text{ are integers} \tag{eqn. A.1}$$

It follows that, if a and b are coprime, so that $h = 1$

$$1 = \lambda a + \mu b \tag{eqn. A.2}$$

8 Groups

8.1 Binary Operations

In Chapter 1 of Volume 1, we considered the combination of real numbers under the binary operations of addition, subtraction, multiplication and division. It is useful and fruitful to generalize the concept of a binary operation by which two elements of any set are combined to form a third element in accordance with some prescribed rule.

DEFINITION 8.1

A *binary operation*, denoted by the symbol $\circ$, is a rule by which two elements x, y of a set S are combined to form a third element $z = x \circ y$. (In a generalized sense, $x \circ y$ will often be called the "product" of x and y under the binary operation $\circ$.)

(*a*) *Closure*

A *closed* binary operation $\circ$ in S is such that $x \circ y$ is defined and is in S for all *ordered* pairs x, y.

(The word "ordered" is included because $x \circ y$ is not necessarily the same as $y \circ x$.)

(*b*) *Commutative Binary Operations*

A *commutative* binary operation $\circ$ in a set S is such that if $x \circ y$ is defined so is $y \circ x$ and $x \circ y = y \circ x$.

(*c*) *Associative Binary Operations*

If the operation $\circ$ is such that whenever one of the two elements $(x \circ y) \circ$

and $x \circ (y \circ z)$ is defined so is the other and

$$(x \circ y) \circ z = x \circ (y \circ z)$$

then $\circ$ is an associative binary operation.

For example, in the set of real numbers, addition and multiplication are associative operations but subtraction and division are not.

(d) Neutral or Identity Elements
Let e be an element of S such that, whenever $x \circ e$ is defined,

$$x \circ e = e \circ x = x$$

then e is called a *neutral* or *identity* element of S for the operation $\circ$.

For example, in the set of real numbers, 1 is the identity element for the operation of multiplication and 0 for addition.

(e) Inverse Elements
Let $\circ$ be a closed binary operation in S and let e be the corresponding identity element. If $x, x' \in S$ are such that

$$x \circ x' = x' \circ x = e$$

then x' is the inverse of x and x is the inverse of x'.

For example, in the set of real numbers, every real number x has an inverse $-x$ under addition and, with the exception of 0, has an inverse $1/x$ or x^{-1} under multiplication.

8.2 *The Nature of Groups*

Amongst the mathematical systems comprising sets, which are subject to only *one* binary operation, the most important are those known as *groups*. Since the laws applicable to groups are few and simple (see Definition 8.2 below) many structures are groups. It frequently happens that several apparently unrelated systems have the same group structure. Consequently, a study of the group structure of one particular system will provide an analysis of the properties of each of the systems.

DEFINITION 8.2
A group is a set S, subject to a binary operation $\circ$, and satisfying the conditions:

(i) *Closure:* for every $x, y \in S$,

$$x \circ y \in S$$

(ii) *Associativity:* for every $x, y, z \in S$,

$$(x \circ y) \circ z = x \circ (y \circ z)$$

(iii) *Identity or Neutral Element:* there is a unique element $e \in S$ such that for every $x \in S$,

$$x \circ e = x = e \circ x$$

(iv) *Inverses:* for every $x \in S$, there is an element $x^{-1} \in S$ such that

$$x \circ x^{-1} = e = x^{-1} \circ x$$

A group which obeys the *commutative* law for products so that for every $x, y \in S$

$$x \circ y = y \circ x$$

is called an *Abelian group.**

EXAMPLES 8.1

(A) The set of integers is a group with respect to addition. The set of positive integers is *not* a group with respect to addition since it has no identity element nor is it a group under multiplication since it has no inverses.

The set of positive integers and their reciprocals do *not* form a group under multiplication since products are not always in the set, e.g.

$$5 \times \tfrac{1}{6} = \tfrac{5}{6} \quad \text{which is not in the set}$$

(B) The set of complex numbers forms a group under addition and multiplication.

(C) The set of residues (mod 5), excluding 0, form a group under multiplication.

* N. H. Abel, the Norwegian mathematician (1802–29), was a pioneer in group theory.

×	1	2	3	4
1	1	2	3	4
2	2	4	1	3
3	3	1	4	2
4	4	3	2	1

TABLE 8.1

The multiplication Table 8.1 for the set of residues 1, 2, 3, 4 (mod 5) shows that they form a group under multiplication since

(i) the product of every pair exists, is unique and is in the set;
(ii) the associative law holds, e.g. $(2 . 4) . 3 = 3 . 3 = 4$ and $2 . (4 . 3) = 2 \cdot 2 = 4$;
(iii) the neutral element is 1;
(iv) every element has a unique inverse: the inverse of a is the unique solution of $ax = 1$.

The group is Abelian.

D) The set of residues (mod 4) under addition.

The addition Table 8.2 for the set residues 0, 1, 2, 3 (mod 4) shows that they

+	0	1	2	3
0	0	1	2	3
1	1	2	3	0
2	2	3	0	1
3	3	0	1	2

TABLE 8.2

form a group (Abelian) under addition with a neutral element 0 and the inverse of any residue x is $(4 - x)$.

E) Let the rectangular axes Ox, Oy be rotated about O through an angle α $(-\pi < \alpha \leqslant \pi)$ in the anticlockwise sense into the positions Ox', Oy' (Fig. 8.1).

Let r_α denote this rotation; it will be equivalent to the following transformation of coordinates:

$$x' = x \cos \alpha + y \sin \alpha \qquad y' = -x \sin \alpha + y \cos \alpha$$

Let r_α, r_β be a pair of transformations and let their product $r_\beta r_\alpha$ denote the transformation resulting from a rotation through an angle α followed by a

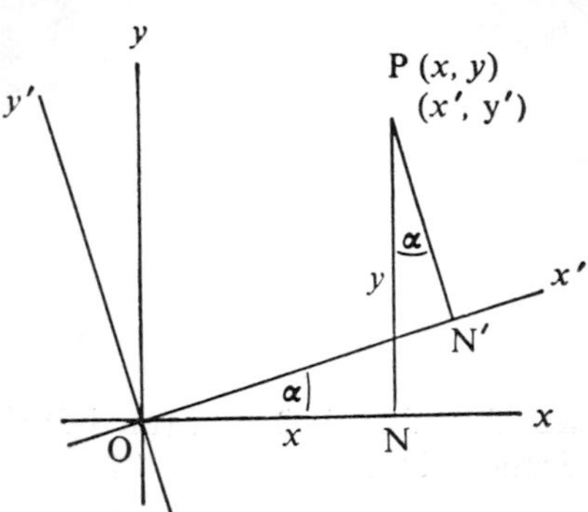

Figure 8.1

rotation through an angle β. Thus, $r_\beta r_\alpha$ defines a closed binary operation in the set of rotations about O. The set of all transformations r_α form an Abelian group, containing an infinite number of elements, for which the neutral element is r_0 and the inverse of r_α is $r_{-\alpha}$.

(F) With reference to Chapter 9 of Volume 1 (Vectors and Matrices), the set comprising the following eight simple transformations forms a non-Abelian group under multiplication.

$$I = \begin{pmatrix} 1 & 0 \\ 0 & 1 \end{pmatrix} \quad \text{Identity}$$

$$A = \begin{pmatrix} 1 & 0 \\ 0 & -1 \end{pmatrix} \quad \text{Reflection in O}x$$

$$B = \begin{pmatrix} -1 & 0 \\ 0 & 1 \end{pmatrix} \quad \text{Reflection in O}y$$

$$C = \begin{pmatrix} 0 & 1 \\ 1 & 0 \end{pmatrix} \quad \text{Reflection in the line } y = x$$

$$D = \begin{pmatrix} 0 & -1 \\ -1 & 0 \end{pmatrix} \quad \text{Reflection in the line } y = -x$$

$$E = \begin{pmatrix} 0 & -1 \\ 1 & 0 \end{pmatrix} \quad \text{Quarter turn}$$

$$F = \begin{pmatrix} -1 & 0 \\ 0 & -1 \end{pmatrix} \quad \text{Half turn}$$

$$G = \begin{pmatrix} 0 & 1 \\ -1 & 0 \end{pmatrix} \quad \text{Three-quarter turn}$$

Under the operation of multiplication, any pair of the above matrices leads to another member of the set (closure). For example, a half turn followed by a reflection in Oy is equivalent to a reflection in Ox:

$$BF = \begin{pmatrix} -1 & 0 \\ 0 & 1 \end{pmatrix}\begin{pmatrix} -1 & 0 \\ 0 & -1 \end{pmatrix} = \begin{pmatrix} 1 & 0 \\ 0 & -1 \end{pmatrix} = A$$

It may easily be verified that $FB = A$ also, so that $BF = FB = A$. In general, products are however *non-commutative*, for example $CB = G$, whilst $BC = E$. Table 8.3 shows the results of multiplying pairs of elements of the set (element

×	*I*	*A*	*B*	*C*	*D*	*E*	*F*	*G*
I	*I*	*A*	*B*	*C*	*D*	*E*	*F*	*G*
A	*A*	*I*	*F*	*G*	*E*	*D*	*B*	*C*
B	*B*	*F*	*I*	*E*	*G*	*C*	*A*	*D*
C	*C*	*E*	*G*	*I*	*F*	*A*	*D*	*B*
D	*D*	*G*	*E*	*F*	*I*	*B*	*C*	*A*
E	*E*	*C*	*D*	*B*	*A*	*F*	*G*	*I*
F	*F*	*B*	*A*	*D*	*C*	*G*	*I*	*E*
G	*G*	*D*	*C*	*A*	*B*	*I*	*E*	*F*

TABLE 8.3

of row postmultiplied by element of column). It will be observed that each of the eight elements occurs once only in each row and column of Table 8.3 and that there is an identity element I.

The operation of multiplication is *associative*, for example

$$C(BD) = CG = B \qquad (CB)D = GD = B$$

i.e. $C(BD) = (CB)D$. This will be found to be true for all triplets of elements. Every element has an *inverse* which is identified by the position of I in the multiplication table. Elements I, A, B, C, D and F are their own inverses which is otherwise obvious on geometrical grounds. Also $EG = I = GE$.

The set of eight transformations is thus a non-Abelian group. It will be observed that the subset I, E, F, G also form a group under multiplication, the product table being as in Table 8.4. It is said to be a *subgroup* of the main group. Subgroups will be considered further in Section 8.12.

×	I	E	F	G
I	I	E	F	G
E	E	F	G	I
F	F	G	I	E
G	G	I	E	F

TABLE 8.4

(G) Table 8.5 defines a binary operation $\circ$ between four elements e, a, b, c comprising a set. It may easily be verified that e, a, b, c form a group. This group of four elements is known as the *Vierergruppe*. Such a set of abstract symbols, having a group structure, is called an *abstract group* which serves as a mathematical model from which general deductions may be made which are applicable to particular sets having this group structure in common.

	e	a	b	c
e	e	a	b	c
a	a	e	c	b
b	b	c	e	a
c	c	b	a	e

TABLE 8.5

(H) With reference to Section 7.5, it is clear that residues (mod 6) excluding 0 do *not* form a group under multiplication since $x = 2$, 3 and 4 have no inverses whilst $2 . 3 = 3 . 2 = 0$ and $3 . 4 = 4 . 3 = 0$ and 0 is not in the set. We have seen that $rx \equiv 1$ (n) has a solution only when r and n are coprime. Residues for a prime modulus will form a group under multiplication if 0 is excluded.

8.3 The Order of a Group

Examples (C), (D), (F), (G) above concern groups each of which contains a finite number of elements; such groups are known as *finite* groups. A "product" table may be produced for any finite group. The number of distinct elements in a finite group is called the order of that group, e.g. the order of the Vierergruppe (G) is 4.

Groups such as those in Examples (B), (E), which are not finite groups are said to be of *infinite* order.

8.4 *Notation*

The following notation will be used throughout the remainder of this chapter: G will denote a general group and e its identity element. The product of elements x, y will be denoted by xy and the inverse of an element x by x^{-1}.

8.5 *The Inverse of a Product*

To PROVE that $(xy)^{-1} = y^{-1}x^{-1}$. Since the associative law $(xy)z = x(yz)$ applies to the group,

$$(y^{-1}x^{-1})(xy) = y^{-1}[x^{-1}(xy)] = y^{-1}[(x^{-1}x)y] = y^{-1}ey = y^{-1}y = e$$

and

$$(xy)(y^{-1}x^{-1}) = x[y(y^{-1}x^{-1})] = x[(yy^{-1})x^{-1}] = xex^{-1} = xx^{-1} = e$$

It follows that $y^{-1}x^{-1}$ is the inverse of xy.

8.6 *The Index Laws*

We define $x^2 = xx$, $x^3 = x^2x = xxx$, $x^4 = x^3x, \ldots, x^n = x^{n-1}x$. Since the associative law holds, it is fairly obvious that the usual index laws

$$x^m x^n = x^{m+n} \quad \text{and} \quad (x^m)^n = x^{mn}$$

also hold for m, n positive integers.

Also, since by the associative law

$$x^n(x^{-1})^n = e$$

it follows that $(x^{-1})^n$ is the inverse of x^n, i.e. $(x^n)^{-1} = (x^{-1})^n$.

The above laws are readily extended to apply to cases where m, n are positive or negative integers. As in elementary algebra, we define x^0 to be e and x^{-n} to be $(x^{-1})^n$, $n > 0$.

8.7 *Isomorphism*

If, in Table 8.1 (p. 227) which is the multiplication table for residues (mod 5), the abstract symbols e, a, c, b are substituted for 1, 2, 3, 4 respectively, we obtain the abstract group table shown in Table 8.6. By interchanging the order of b and c, this table becomes as Table 8.7. Again, if in Table 8.2 (p. 227) which is the addition table for residues (mod 4), the symbols e, a, b, c are

	e	*a*	*c*	*b*
e	*e*	*a*	*c*	*b*
a	*a*	*b*	*e*	*c*
c	*c*	*e*	*b*	*a*
b	*b*	*c*	*a*	*e*

TABLE 8.6

substituted for 0, 1, 2, 3 respectively, we obtain the abstract group Table 8.7 once again.

	e	*a*	*b*	*c*
e	*e*	*a*	*b*	*c*
a	*a*	*b*	*c*	*e*
b	*b*	*c*	*e*	*a*
c	*c*	*e*	*a*	*b*

TABLE 8.7

Hence, if the one-to-one correspondences in Table 8.8 are used, the group structures of the residues (mod 5) under multiplication and of the residues

	× mod 5		+ mod 4
e →	1	→	0
a →	2	→	1
b →	4	→	2
c →	3	→	3

TABLE 8.8

(mod 4) under addition are identical. Two such groups are said to be *isomorphic.*

Infinite groups, as well as finite groups, exhibit isomorphism. For example, when both are under addition the group of integers (positive, negative and zero) is isomorphic to the group of even integers, with the following one-to-one correspondence

$$\ldots -3, -2, -1, 0, 1, 2, 3, \ldots, n, \ldots$$
$$\ldots -6, -4, -2, 0, 2, 4, 6, \ldots, 2n, \ldots$$

The "products" of two elements from each group exhibit the same correspondence:

$(m + n)$ from the first group → $(2m + 2n)$ from the second

EXERCISES 8.1

1. State which of the following sets are groups for the specified binary operations. For those which are groups, give the identity element and some examples of the inverses of the other elements. Give reasons why the remaining sets are not groups.

(*a*) *Sets of numbers*

(i) Positive integers (excluding 0) under addition.
(ii) Even integers (positive, negative and zero) under addition.
(iii) All integers (positive, negative and zero) under multiplication.
(iv) Odd integers under addition.
(v) Real numbers under multiplication.
(vi) Real numbers under division.
(vii) Rational numbers under multiplication.
(viii) Rational numbers, excluding 0, under multiplication.
(ix) All integers (positive, negative and zero) divisible by 3 under addition.
(x) All numbers of the form 3^n (with n a positive or negative integer or zero) under multiplication.

(*b*) *Sets of residues*

(xi) Residues, mod 4, under multiplication.
(xii) Residues (0, 2, 4, 6), mod 7, under addition.
(xiii) Residues excluding 0, mod 7, under multiplication.
(xiv) Residues, mod 8, under addition.
(xv) Residues excluding 0, mod 8, under multiplication.
(xvi) Residues (1, 2, 4, 5, 7, 8), mod 9, under multiplication.
(xvii) Residues (1, 3, 4, 5, 9), mod 11, under multiplication.

[Nos. i, iii, iv, vi, xi, xii and xv are *not* groups for the reasons given below.
(i) No neutral element. (iii) No inverses. (iv) No identity element (unless 0 included). Closure requirements unsatisfied. (vi) Associativity requirements unsatisfied. (xi) No inverses for 0 and 2. (xii) and (xv) Closure and inverse requirements unsatisfied.]

2. Show that the set of residues excluding 0, mod 12, does not form a group under multiplication. Show further that subset (1, 3, 5, 7, 9, 11) is not a group but that the subset (1, 5, 7, 11) is a group. What are the inverses of the elements of (1, 5, 7, 11)? [Inverses are (1, 5, 7, 11)]

3. Complete the table on p. 234 for multiplication modulo 8: where the entry in the rth row and sth column is the remainder on dividing the ordinary arithmetical product rs by 8.

Find the largest subset of the numbers 1, 2, 3, . . . , 7 which forms a group under this rule of combination.

Prove that, if a, b, x are elements of the subset just obtained, and if $ax \equiv b \pmod 8$, then

$$x \equiv ab \pmod 8$$

(*Oxford and Cambridge G.C.E. A-level, S.M.P.*)

×	1	2	3	4	5	6	7
1	1	2	3	4	5	6	7
2	2	4	6	0			
3	3	6	1				
4	4	0	4				
5	5	2					
6	6	4					
7	7						

4. Give reasons why the following sets are groups:
   (i) Vectors of the form (kx, x), where k is a scalar constant, under vector addition.
   (ii) Non-singular (2×2) matrices under multiplication.

8.8 Permutation Groups

Let n different objects be labelled 1, 2, 3, . . . , n. These objects may be arranged amongst themselves in $n!$ different ways, i.e. there are $n!$ *permutations*. Each of these permutations of n objects may be represented by a particular order of the numbers 1, 2, 3, . . . , n, and is defined to be a permutation of degree n.

A permutation of degree n may conveniently be written in the form

$$\begin{pmatrix} 1 & 2 & 3 & \cdots & i & \cdots & n \\ a_1 & a_2 & a_3 & \cdots & a_i & \cdots & a \end{pmatrix}$$

which indicates that the numbers 1, 2, 3, . . . , n are replaced by a different order of themselves, namely by $a_1, a_2, a_3, \ldots, a_i, \ldots, a_n$ respectively, i.e. i is replaced by a_i for all values of i. For example

$$\begin{pmatrix} 1 & 2 & 3 & 4 \\ 3 & 1 & 4 & 2 \end{pmatrix}$$

represents the permutation of 1, 2, 3, 4 changing 1 into 3, 2 into 1, 3 into 4 and 4 into 2.

The 6 possible permutations of 3 objects are

$$\begin{pmatrix} 1 & 2 & 3 \\ 1 & 2 & 3 \end{pmatrix} \quad \begin{pmatrix} 1 & 2 & 3 \\ 1 & 3 & 2 \end{pmatrix} \quad \begin{pmatrix} 1 & 2 & 3 \\ 2 & 1 & 3 \end{pmatrix} \quad \begin{pmatrix} 1 & 2 & 3 \\ 2 & 3 & 1 \end{pmatrix}$$

$$\begin{pmatrix} 1 & 2 & 3 \\ 3 & 1 & 2 \end{pmatrix} \quad \begin{pmatrix} 1 & 2 & 3 \\ 3 & 2 & 1 \end{pmatrix}$$

Note that the order of the numbers in the upper row is not significant since we are only concerned with the changes which take place so that, for example

$$\begin{pmatrix}1 & 2 & 3\\2 & 1 & 3\end{pmatrix} \qquad \begin{pmatrix}2 & 3 & 1\\1 & 3 & 2\end{pmatrix} \qquad \begin{pmatrix}1 & 3 & 2\\2 & 3 & 1\end{pmatrix}$$

all represent the same permutation. It is convenient to denote the permutations

$$\begin{pmatrix}1 & 2 & 3 & \cdots & i & \cdots & n\\a_1 & a_2 & a_3 & \cdots & a_i & \cdots & a_n\end{pmatrix} \text{ by } \begin{pmatrix}i\\a_i\end{pmatrix}$$

where i takes the values $1, 2, 3, \ldots, n$ successively.

PRODUCT OF TWO PERMUTATIONS OF THE SAME *n* OBJECTS

Let $x = \begin{pmatrix}i\\a_i\end{pmatrix}$ and $y = \begin{pmatrix}i\\b_i\end{pmatrix}$

The product xy is defined to be the permutation which results when $1, 2, 3, \ldots, n$ are first rearranged in accordance with the permutation y and the resulting integers are again rearranged in accordance with the permutation x. For example, if

$$x = \begin{pmatrix}1 & 2 & 3 & 4\\4 & 3 & 1 & 2\end{pmatrix} \quad \text{and} \quad y = \begin{pmatrix}1 & 2 & 3 & 4\\3 & 2 & 4 & 1\end{pmatrix}$$

then*

$$\begin{aligned} xy &= \begin{pmatrix}1 & 2 & 3 & 4\\4 & 3 & 1 & 2\end{pmatrix}\begin{pmatrix}1 & 2 & 3 & 4\\3 & 2 & 4 & 1\end{pmatrix}\\ &= \begin{pmatrix}3 & 2 & 4 & 1\\1 & 3 & 2 & 4\end{pmatrix}\begin{pmatrix}1 & 2 & 3 & 4\\3 & 2 & 4 & 1\end{pmatrix}\\ &= \begin{pmatrix}1 & 2 & 3 & 4\\1 & 3 & 2 & 4\end{pmatrix}\end{aligned}$$

* For convenience, the permutation x has been rewritten with the order of its first row identical to the order of the second row of y.

Note that

$$yx = \begin{pmatrix} 1 & 2 & 3 & 4 \\ 3 & 2 & 4 & 1 \end{pmatrix}\begin{pmatrix} 1 & 2 & 3 & 4 \\ 4 & 3 & 1 & 2 \end{pmatrix}$$

$$= \begin{pmatrix} 4 & 3 & 1 & 2 \\ 1 & 4 & 3 & 2 \end{pmatrix}\begin{pmatrix} 1 & 2 & 3 & 4 \\ 4 & 3 & 1 & 2 \end{pmatrix}$$

$$= \begin{pmatrix} 1 & 2 & 3 & 4 \\ 1 & 4 & 3 & 2 \end{pmatrix}$$

so that $xy \neq yx$. Products are *not*, in general, commutative. In the general case, we may write

$$xy = \begin{pmatrix} i \\ a_i \end{pmatrix}\begin{pmatrix} i \\ b_i \end{pmatrix} = \begin{pmatrix} b_i \\ a_{b_i} \end{pmatrix}\begin{pmatrix} i \\ b_i \end{pmatrix} = \begin{pmatrix} i \\ a_{b_i} \end{pmatrix}$$

whilst

$$yx = \begin{pmatrix} i \\ b_{a_i} \end{pmatrix}$$

SYMMETRIC GROUP OF DEGREE *n*

It may be shown that the set of all permutations $\begin{pmatrix} i \\ a_i \end{pmatrix}$ $(i = 1, 2, \ldots, n)$ containing $n!$ elements, forms a group under multiplication.

The product of any two permutations x, y is unique and is in the set. It may readily be verified that the associative law holds for products.

The identity element is $\begin{pmatrix} i \\ i \end{pmatrix} = e$. For example,

$$ex = \begin{pmatrix} i \\ i \end{pmatrix}\begin{pmatrix} i \\ a_i \end{pmatrix} = \begin{pmatrix} a_i \\ a_i \end{pmatrix}\begin{pmatrix} i \\ a_i \end{pmatrix} = \begin{pmatrix} i \\ a_i \end{pmatrix} = x$$

$$xe = \begin{pmatrix} i \\ a \end{pmatrix}\begin{pmatrix} i \\ i \end{pmatrix} = \begin{pmatrix} i \\ a_i \end{pmatrix} = x$$

Every element x has an inverse. The inverse of $\begin{pmatrix} i \\ a_i \end{pmatrix}$ is $\begin{pmatrix} a_i \\ i \end{pmatrix}$ and we have

$$xx^{-1} = \begin{pmatrix} i \\ a_i \end{pmatrix}\begin{pmatrix} a_i \\ i \end{pmatrix} = \begin{pmatrix} a_i \\ a_i \end{pmatrix} = e$$

$$x^{-1}x = \begin{pmatrix} a_i \\ i \end{pmatrix}\begin{pmatrix} i \\ a_i \end{pmatrix} = \begin{pmatrix} i \\ i \end{pmatrix} = e$$

It follows that this set of permutations is a non-Abelian group of order $n!$ under multiplication provided $n \geqslant 3$ and is called the *symmetric group* of degree n and will be denoted by S_n.

8.9 The Order of an Element of a Group

In the group Table 8.4 (p. 230) for the group (I, E, F, G), each element of which is a (2×2) matrix, let us find the powers of the various elements. All powers of the identity element I are equal to I.

It is easily verified that

$$E^1 = E \qquad E^2 = F \qquad E^3 = EF = FE = G \qquad E^4 = EG = GE = I$$

$$E^5 = EI = E \qquad E^6 = IE^2 = F$$

The first four powers of E are E, F, G, I, the four elements of the group and, since $E^4 = I$, higher powers of E must necessarily repeat the values E, F, G, I.

Similarly, we find

$$F^1 = F \qquad F^2 = I \qquad F^3 = IF = F \qquad F^4 = I^2 = I \quad \text{and so on}$$

$$G^1 = G \qquad G^2 = F \qquad G^3 = GF = FG = E$$

$$G^4 = GE = EG = I \qquad G^5 = IG = G \quad \text{and so on}$$

In the general case, let x be any element of a group. Then, by the closure property, the "product" xx, i.e. x^2, is an element of the group (here "product" is used, in a general sense, to denote the result of applying the particular binary operation of the group to a pair of elements). In fact, $x, x^2, x^3, \ldots,$ are all elements of the group.

In an infinite group, it is possible that all these elements may be different. However, there are, at most, n different elements in a group of order n so that ultimately two of these powers of x must be identical.

Thus for some p, q $(q > p)$, $x^q = x^p$

Therefore $x^{q-p} = e$ i.e. $x^k = e$ $(k > 0)$

Thus, in a finite group, there are positive values of k such that $x^k = e$. The smallest value of k for which $x^k = e$ is called the *order* (or period) of the element x.

It follows that, in the example at the beginning of this section, E, G are elements of order 4 (the order of the group), F is of order 2, and the identity element I is of order 1, as it is in any group.

EXAMPLE 8.2

(A) Consider the residues (mod 12), excluding 0, under multiplication. Only the residues which are coprime with 12, i.e. 1, 5, 7, 11, form a group (see Exercises 8.1, No. 2).

The group table for these residues is given in Table 8.9.

×	1	5	7	11
1	1	5	7	11
5	5	1	11	7
7	7	11	1	5
11	11	7	5	1

TABLE 8.9

The powers of the various elements are

Element		*Order*
1		1
5	$5^2 = 1$	2
7	$7^2 = 1$	2
11	$11^2 = 1$	2

(B) Consider the residues (mod 7), excluding 0, under multiplication. Since 7 is prime, they form a group for which the multiplication table is as shown in Table 8.10.

×	1	2	3	4	5	6
1	1	2	3	4	5	6
2	2	4	6	1	3	5
3	3	6	2	5	1	4
4	4	1	5	2	6	3
5	5	3	1	6	4	2
6	6	5	4	3	2	1

TABLE 8.10

The powers of the various elements are

Element		*Order*
1		1
2	$2^2 \equiv 4,\ 2^3 \equiv 2.4 \equiv 1$	3
3	$3^2 \equiv 2,\ 3^3 \equiv 3.2 \equiv 6,\ 3^4 \equiv 3.6 \equiv 4,$ $3^5 \equiv 3.4 \equiv 5,\ 3^6 \equiv 3.5 \equiv 1$	6
4	$4^2 \equiv 2,\ 4^3 \equiv 4.2 \equiv 1$	3
5	$5^2 \equiv 4,\ 5^3 \equiv 5.4 \equiv 6,\ 5^4 \equiv 5.6 \equiv 2,$ $5^5 \equiv 5.2 \equiv 3,\ 5^6 \equiv 5.3 \equiv 1$	6
6	$6^2 \equiv 1$	2

8.10 Cyclic Groups

THEOREM 8.1
If the element x of a group has order r, then $e, x, x^2, \ldots, x^{r-1}$ are r distinct elements of the group.

Suppose they are not all distinct, and let $x^m = x^n$ $(r - 1 \geqslant m > n \geqslant 0)$. Therefore

$$x^{m-n} = e \qquad \text{i.e. } x^t = e \qquad t = m - n > 0 \quad (0 < t < r - 1)$$

which is contrary to the hypothesis that the order of x is r. Thus the first r powers of x are distinct.

Also $x^{r+1} = x, x^{r+2} = x^2, x^{r+3} = x^3, \ldots$, so that the elements are repeated indefinitely for higher powers of x.

$$x^p = e \text{ implies that } p \text{ is a multiple of } r$$

Since a group of order n contains n distinct elements, no element can have an order greater than n. An element of order n is such that $x^n = e$, and $e, x, x^2, \ldots, x^{n-1}$ are n distinct elements of the group and therefore comprise the whole group. Thus x generates the whole group. Such a group is said to be *cyclic* and x is a *generator* of the group. A group may have several generators. A cyclic group of order n will be denoted by C_n.

With reference to Example 8.2(B), it is seen that the residues 3, 5 are generators of the group of residues (mod 7), excluding 0, under multiplication and that they are the only generators.

If x is an element of order r $(<n)$ of a group of order n, then $e, x, x^2, \ldots, x^{r-1}$ is a (cyclic) subgroup of the main group. For example, in the group of residues (mod 7) (Example 8.2(B)), the residue 2 is of order 3 since $2^3 \equiv 1$ (7). We have, $2^1 = 2,\ 2^2 = 4,\ 2^3 = 1$ (mod 7) so that the cyclic

group generated by 2 is 1, 2, 4. This is a subgroup of the whole group of residues (mod 7) and has the multiplication table shown in Table 8.11.

×	1	2	4
1	1	2	4
2	2	4	1
4	4	1	2

TABLE 8.11

THEOREM 8.2
If x is a generator of a cyclic group of order n, then x^r is of order n provided r is prime to n.

Since x is a generator, then $x^{\lambda n} = e$, where λ is an integer.
Let $(x^r)^s = e$, then $rs = \lambda n$ (a multiple of n).
Since r is prime to n, s must be a multiple of n, and the smallest value of s such that $(x^r)^s = e$ is $s = n$.
Therefore $(x^r)^n = e$ and x^r is a generator of the group.

THEOREM 8.3
If r is *not* prime to n and l is the l.c.m. of r and n, then x^r is of order l/r.

The smallest value of s such that rs is a multiple of n is given by $rs = l$, i.e. $s = l/r$. Therefore the order of x^r is l/r.

These theorems may be exemplified from the cyclic group of order 6:

$$e, x, x^2, \ldots, x^5, x^6 = e$$

$r = 1, 5$ (only) are prime to 6. For $r = 1$, we have $x^r = x^1$ and since $x^6 = e$, x is of order 6.
x^5 is of order 6 since the successive powers of x^5 are x^5, $x^{10} = x^4$, $x^{15} = x^3$, $x^{20} = x^2$, $x^{25} = x$, $x^{30} = e$.
$r = 2, 3, 4$ are not prime to 6.

r	*Powers of* x^r	*Order of* x^r	l	l/r
2	$x^2, x^4, x^6 = e$	3	6	3
3	$x^3, x^6 = e$	2	6	2
4	$x^4, x^8 = x^2, x^{12} = e$	3	12	3

As an illustration of a cyclic group of order 6, we find that 3 is a generator of the cyclic group of residues (mod 7), excluding 0, under multiplication. It may readily be verified that (3^5) is also of order 6, whilst (3^2) and (3^4) are of order 3 and (3^3) is of order 2.

INFINITE CYCLIC GROUPS

An infinite group G is said to be cyclic if there is an element x belonging to G such that G consists of

$$\ldots, x^{-r}, \ldots, x^{-2}, x^{-1}, e, x, x^2, \ldots, x^r, \ldots$$

i.e. of all the positive and negative integral powers of x together with the neutral element e. The inverse of x^r is x^{-r}. x is called the generator of the group.

8.11 Transformation Groups

If a rectangle is rotated through 180°: (*a*) in its own plane about its centre in either the clockwise or the anticlockwise sense or (*b*) about a line through its centre parallel to one pair of its sides, the rectangle will occupy the same position though its vertices will be in different positions. Such a movement of the figure is called a *geometrical transformation* or simply a *transformation.*

Many examples of figures with a high degree of symmetry occur for which the set of all such transformations forms a group.

1. THE SYMMETRY GROUP OF THE RECTANGLE

Let r denote the transformation resulting from a clockwise rotation of 180° about O, the centre of the rectangle (Fig. 8.2), and let p denote the transformation resulting from a rotation of 180° about Oy || BC. The product pr

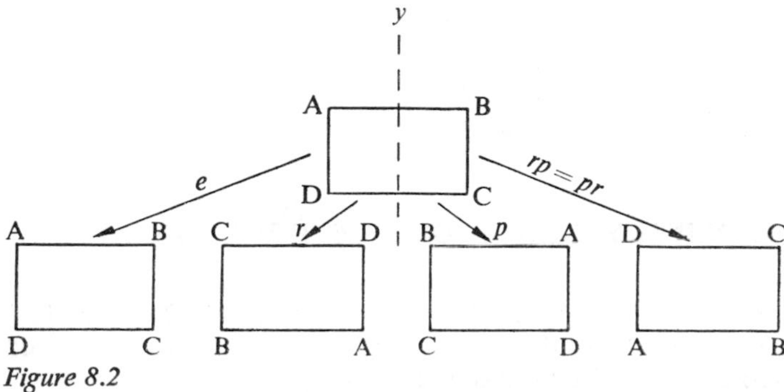

Figure 8.2

will denote the transformation corresponding to a transformation r *followed by* a transformation p.

If r is applied twice in succession, the rectangle will be rotated through $2 \times 180°$ in the clockwise sense, i.e. the transformation r^2 restores the rectangle to its original position. If e denotes the neutral element, i.e. e is the transformation which leaves the rectangle quite unchanged, we have $r^2 = e$. Similarly $p^2 = e$.

At first sight, it might appear that an infinite number of transformations are possible but this is not the case since, for example, r^3 represents a rotation through $3 \times 180°$ which has the same effect as a rotation through 180°, i.e. $r^3 = r$ and $r^4 = e$, $r^5 = r$, and so on.

r^{-1}, the inverse of r, denotes the transformation resulting from an anti-clockwise rotation of 180°.

From Fig. 8.2, it is obvious that $pr = rp$.

The set of four elements (e, r, p, pr) form a group since their product table (Table 8.12) shows that the set is closed under multiplication, each

$\times$	e	r	p	pr
e	e	r	p	pr
r	r	e	pr	p
p	p	pr	e	r
pr	pr	p	r	e

TABLE 8.12

element has an inverse and, as may readily be verified, the associative law holds. This group is an example of the Vierergruppe (Table 8.5, p. 230) with the correspondence $a = r$, $b = p$, $c = pr$. It is also called the *dihedral group* of order 4 ("dihedral" refers to symmetries involving both sides of the rectangle). This group contains four subgroups of order 2 as shown in Table 8.13.

$\times$	e	r
e	e	r
r	r	e

$\times$	e	p
e	e	p
p	p	e

$\times$	e	pr
e	e	pr
pr	pr	e

TABLE 8.13

2. THE SYMMETRY GROUP OF THE EQUILATERAL TRIANGLE

Let the transformations resulting from (i) a rotation of 120° in a clockwise sense about O, the centre of the triangle, and (ii) a half-turn about an altitude AN (Fig. 8.3) be denoted by r and p respectively.

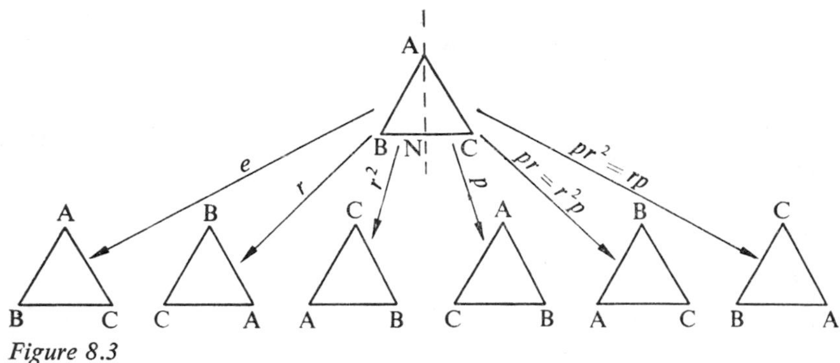

Figure 8.3

r^2 represents a rotation through 240° and r^3 a rotation through 360°. Therefore $r^3 = e$.

The set (e, r, r^2) is a cyclic (rotation) group of order 3.

As in 1 (above), $p^2 = e$.

If any one of the transformations e, r, r^2 is applied and then the triangle is turned about the altitude AN, a new transformation is obtained. Thus, there is a set of six transformations:

$$(e, r, r^2, p, pr, pr^2)$$

It is easily verified from Fig. 8.3 that

$$pr^2 = rp \quad \text{and} \quad pr = r^2p$$

The product table (Table 8.14) shows that the above set of six transformations

×	e	r	r^2	p	pr	pr^2
e	e	r	r^2	p	pr	pr^2
r	r	r^2	e	pr^2	p	pr
r^2	r^2	e	r	pr	pr^2	p
p	p	pr	pr^2	e	r	r^2
pr	pr	pr^2	p	r^2	e	r
pr^2	pr^2	p	pr	r	r^2	e

TABLE 8.14

form a group. It is, in fact, the smallest non-Abelian group, and is called the *dihedral group* of order 6.

(Note that many products in Table 8.14 may be simplified by using the relationships $pr^2 = rp, pr = r^2p$ above, e.g. $pr^2 \,.\, pr^2 = p \,.\, pr \,.\, r^2 = p^2r^3 = e$.)

This group contains a number of subgroups. Apart from the trivial subgroup e, there is one subgroup of order 3, a cyclic (rotation) group, whose product table is at the top left-hand corner of Table 8.14. There are also three subgroups of order 2 as in Table 8.15. It may be shown that the

	e	p
e	e	p
p	p	e

	e	pr
e	e	pr
pr	pr	e

	e	pr^2
e	e	pr^2
pr^2	pr^2	e

TABLE 8.15

symmetric group of degree 3 (see Section 8.8) is isomorphic to the symmetry group of the equilateral triangle. The symmetric group of degree 3 is of order 3!, i.e. 6.

Any transformation of the equilateral triangle may be represented by a permutation of degree 3. For example, consider the transformation r (Fig. 8.4). Let 1, 2, 3 denote the positions of the vertices of the triangle.

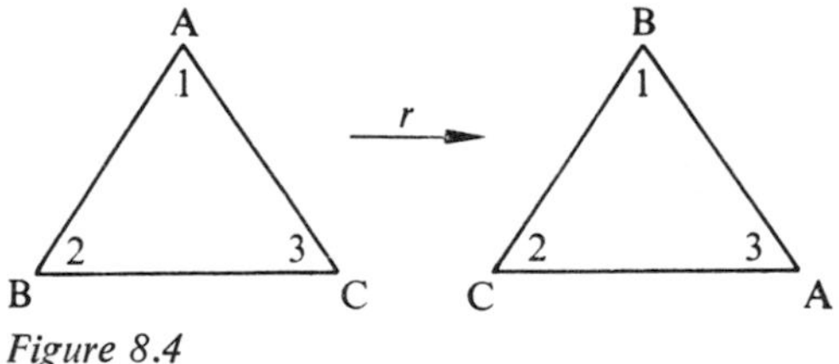

Figure 8.4

The transformation r moves A from 1 to 3, B from 2 to 1, C from 3 to 2, so that the transformation r corresponds to a permutation which will be denoted by r:

$$r = \begin{pmatrix} 1 & 2 & 3 \\ 3 & 1 & 2 \end{pmatrix}$$

For similar reasons, we write

$$p = \begin{pmatrix} 1 & 2 & 3 \\ 1 & 3 & 2 \end{pmatrix} \qquad e = \begin{pmatrix} 1 & 2 & 3 \\ 1 & 2 & 3 \end{pmatrix}$$

$$r^2 = \begin{pmatrix} 1 & 2 & 3 \\ 3 & 1 & 2 \end{pmatrix}\begin{pmatrix} 1 & 2 & 3 \\ 3 & 1 & 2 \end{pmatrix} = \begin{pmatrix} 1 & 2 & 3 \\ 2 & 3 & 1 \end{pmatrix}$$

It may easily be verified from Fig. 8.3 that $\begin{pmatrix}1 & 2 & 3\\ 2 & 3 & 1\end{pmatrix}$ has the same geometrical interpretation as the application of the transformation r twice in succession. Again (r is applied first and then p),

$$pr = \underset{(p)}{\begin{pmatrix}1 & 2 & 3\\ 1 & 3 & 2\end{pmatrix}}\underset{(r)}{\begin{pmatrix}1 & 2 & 3\\ 3 & 1 & 2\end{pmatrix}} = \begin{pmatrix}1 & 2 & 3\\ 2 & 1 & 3\end{pmatrix}$$

It may be verified directly, in this manner, that $pr^2 = rp$ and that $r^2p = pr$.

$$pr^2 = \underset{(p)}{\begin{pmatrix}1 & 2 & 3\\ 1 & 3 & 2\end{pmatrix}}\underset{(r^2)}{\begin{pmatrix}1 & 2 & 3\\ 2 & 3 & 1\end{pmatrix}} = \begin{pmatrix}1 & 2 & 3\\ 3 & 2 & 1\end{pmatrix} = rp$$

$$r^2p = \underset{(r^2)}{\begin{pmatrix}1 & 2 & 3\\ 2 & 3 & 1\end{pmatrix}}\underset{(p)}{\begin{pmatrix}1 & 2 & 3\\ 1 & 3 & 2\end{pmatrix}} = \begin{pmatrix}1 & 2 & 3\\ 2 & 1 & 3\end{pmatrix} = pr$$

It may also be easily verified that $r^3 = p^2 = e$ and that the product of any pair of transformations corresponds to the product of their permutations. Thus there is one-to-one correspondence between the group G of six geometrical transformations (e, r, r^2, p, pr, pr^2) and the six permutations of the symmetry group S_3, i.e. the groups G and S_3 are *isomorphic*.

Alternatively, we may say that if the binary operation in a set of six elements (e, a, b, c, d, f) is defined by Table 8.16 we obtain a group G which

	e	a	b	c	d	f
e	e	a	b	c	d	f
a	a	b	e	f	c	d
b	b	e	a	d	f	c
c	c	d	f	e	a	b
d	d	f	c	b	e	a
f	f	c	d	a	b	e

TABLE 8.16

is identical with the symmetry group S_3 except for the nature of their elements.

3. THE SYMMETRY GROUP OF THE SQUARE

Since the square has a higher degree of symmetry than the rectangle, it is to be expected that this group has a higher order than that of the rectangle. The order of the symmetry group of the square is 8.

Let r denote the transformation resulting from a clockwise rotation of the square about its centre and in its own plane through 90°. Let p denote a half-turn of the square about a line through its centre parallel to a side (or about one of its diagonals). The set of eight transformations $(e, r, r^2, r^3, p, pr, pr^2, pr^3)$, with

$$r^4 = e \qquad p^2 = e \qquad rp = pr^3 \qquad r^2p = pr^2 \qquad r^3p = pr$$

form a group with product table as in Table 8.17.

	e	r	r^2	r^3	p	pr	pr^2	pr^3
e	e	r	r^2	r^3	p	pr	pr^2	pr^3
r	r	r^2	r^3	e	pr^3	p	pr	pr^2
r^2	r^2	r^3	e	r	pr^2	pr^3	p	pr
r^3	r^3	e	r	r^2	pr	pr^2	pr^3	p
p	p	pr	pr^2	pr^3	e	r	r^2	r^3
pr	pr	pr^2	pr^3	p	r^3	e	r	r^2
pr^2	pr^2	pr^3	p	pr	r^2	r^3	e	r
pr^3	pr^3	p	pr	pr^2	r	r^2	r^3	e

TABLE 8.17

There is a cyclic (rotational) subgroup of order 4 whose product table is at the top left-hand corner of Table 8.17.

EXERCISES 8.2

1. Show that the residues, mod 5 and mod 8, form cyclic groups under addition. Find the orders of all their elements and identify their generators. [Mod 5: 1, 2, 3 and 4 are generators; mod 8: 1, 3, 5 and 7 are generators; 2 (order 4), 4 (order 2), 6 (order 4)]
2. Identify the generators of the cyclic groups

   (*a*) (e, x, x^2, x^3, x^4) (*b*) $(e, y, y^2, y^3, y^4, y^5)$.

   [(*a*) x, x^2, x^3, x^4; (*b*) y, y^5]

3. Show that the residues, mod 5, excluding 0, form a cyclic group of order 4 under multiplication, and identify its generators. If x is a generator, illustrate Theorems 8.2 and 8.3 by determining the orders of x and x^3. [Generators are 2 and 3]
4. Show that the residues (1, 3, 5, 7), mod 8, form a group under multiplication. Find the orders of its elements and hence show that it is not isomorphic to C_4 (*see* Section 8.10). Show that it is isomorphic to the Vierergruppe shown in Table 8.5 (p. 230) and specify the correspondence.
[3, 5 and 7, each of order 2; Vierergruppe $a = 3$, $b = 5$, $c = 7$]
5. Show that the residues (1, 3, 7, 9), mod 10, form a cyclic group of order 4 under multiplication. What are the orders and the inverses of its elements? Identify its generators. Show that it is isomorphic to the group (0, 1, 2, 3), mod 4, under addition with the one-to-one correspondence (1, 3, 7, 9) → (0, 1, 3, 2) respectively. State another one-to-one correspondence. Verify that corresponding elements have the same order.
[Generators 3, 7; 9 (order 2); 3 is inverse of 7 and vice versa, 9 is inverse of 9; another correspondence (1, 3, 7, 9) → (0, 3, 1, 2)]
6. Show that the group of residues, mod 7, excluding 0, under multiplication is isomorphic to the group of residues, mod 6, under addition and specify a one-to-one correspondence. [(1, 3, 2, 6, 4, 5) → (0, 1, 2, 3, 4, 5)]
7. Show that the residues, mod 13, excluding 0, form a group under multiplication which is isomorphic to C_{12}. Identify the generators. [Generator is 2]
8. Show that the residues (1, 5, 7, 11), mod 12, form a group isomorphic with the Vierergruppe (Table 8.5) under multiplication. Identify the subgroups.
[(e, a, b, c) → (1, 5, 7, 11)]
9. A binary operation $\circ$ is defined on the set P of all ordered pairs (g, h), where g is any element of a group G and h any element of a group H, by setting

$$(g_1, h_1) \circ (g_2, h_2) = (g_1 g_2, h_1 h_2)$$

for any g_1, g_2 in G and h_1, h_2 in H. Prove that $(P, \circ)$ is a group, i.e. show that the operation $\circ$ is associative, and prove that (e_G, e_H) is a unit element where e_G and e_H are the unit elements of G and H respectively. Find an inverse element for any element (g, h) of P, with respect to the operation $\circ$, and prove that P contains a subgroup isomorphic with G and a subgroup isomorphic with H.

If G is the cyclic group $(1, x, x^2 = 1)$ of order 2, and H is the cyclic group $(1, y, y^2, y^3 = 1)$ of order 3, prove that P is a cyclic group. Find the order of P in this case. [Inverse element (g^{-1}, h^{-1}); generator (x, y), order 6]

10. Show that the cross-ratios $t, \dfrac{1}{t}, 1 - t, \dfrac{1}{1-t}, \dfrac{t-1}{t}, \dfrac{t}{t-1}$

form a group where the operation is the substitution of the second factor for t in the first factor, e.g.

$$\frac{t-1}{t} \circ \frac{1}{t} = \frac{1/t - 1}{1/t} = 1 - t \qquad \text{(another element)}$$

Show that the group is isomorphic to the symmetry group of degree 3, i.e. to the dihedral group of order 6.

11. The elements 0, 1, 2 form a group under addition (mod 3) and the elements 1, $\frac{1}{2}(-1 \pm i\sqrt{3})$ form a group under multiplication. Show that the groups are isomorphic and specify the correspondence. [(0, 1, 2) $\rightarrow 1, \omega, \omega^2$) where $\omega = e^{i2/3}$]

12. Show that, under the one-to-one correspondence $(1, i, -1, -i) \rightarrow (0, 1, 2, 3)$ respectively, the group $(1, i, -1, -i)$ under multiplication is isomorphic to the group of residues (0, 1, 2, 3) mod 4 under addition and also isomorphic to C_4. State another isomorphic correspondence. [$(1, i, -1, -i) \rightarrow (1, 2, 3, 0)$]

13. Complete the table of products for a group consisting of four elements, e, a, b, c. If the elements of the group are taken from the set of complex numbers and the group operation is the ordinary multiplication of complex numbers, find all possible selections of the numbers e, a, b and c. Prove also that, to within isomorphism, there are only two groups of order 4.

$\times$	e	a	b	c
e	e	a	b	c
a	a	b		
b	b			
c	c			

(*Oxford and Cambridge G.C.E. A-level*)
[$(e, a, b, c) \rightarrow (1, 1, 1, 1), (1, -1, 1, -1), (1, i, -1, -i), (1, -i, -1, i)$]

14. Each of the following is a group of order 4:
(*a*) Residues (1, 5, 7, 11) mod 12 under multiplication.
(*b*) The symmetry group of the rectangle.
(*c*) $(x, -x, 1/x, -1/x)$ where the operation is the substitution of the second element for x in the first element.
Show that these groups are isomorphic.

15. If $x = \begin{pmatrix} 1 & 2 & 3 & 4 \\ 2 & 3 & 4 & 1 \end{pmatrix}$ $\quad y = \begin{pmatrix} 1 & 2 & 3 & 4 \\ 3 & 1 & 2 & 4 \end{pmatrix}$ $\quad z = \begin{pmatrix} 1 & 2 & 3 & 4 \\ 2 & 4 & 1 & 3 \end{pmatrix}$

show that $yxy^{-1} = z$. Find (i) x^2y. (ii) yx^2. (iii) $(xy)^2$. (iv) yxy^3.

$$\left[\text{(i)} \begin{pmatrix} 1 & 2 & 3 & 4 \\ 1 & 3 & 4 & 2 \end{pmatrix}, \text{(ii)} \begin{pmatrix} 1 & 2 & 3 & 4 \\ 2 & 4 & 3 & 1 \end{pmatrix}, \text{(iii)} \begin{pmatrix} 1 & 2 & 3 & 4 \\ 1 & 2 & 3 & 4 \end{pmatrix}, \text{(iv)} \begin{pmatrix} 1 & 2 & 3 & 4 \\ 1 & 2 & 4 & 3 \end{pmatrix}\right]$$

16. Show that the three permutations

$$\begin{pmatrix} 1 & 2 & 3 \\ 1 & 2 & 3 \end{pmatrix} \qquad \begin{pmatrix} 1 & 2 & 3 \\ 2 & 3 & 1 \end{pmatrix} \qquad \begin{pmatrix} 1 & 2 & 3 \\ 3 & 1 & 2 \end{pmatrix}$$

form a cyclic group of order 3 which is a subgroup of the symmetric group of degree 3. Verify by direct multiplication of its elements, that the symmetric group S_3 is isomorphic to the dihedral group of order 6. Is the isomorphism uniquely defined?

17. Prove that the Vierergruppe is isomorphic to the subgroup of S_4 whose elements are

$$\begin{pmatrix}1 & 2 & 3 & 4\\1 & 2 & 3 & 4\end{pmatrix} \quad \begin{pmatrix}1 & 2 & 3 & 4\\3 & 4 & 1 & 2\end{pmatrix} \quad \begin{pmatrix}1 & 2 & 3 & 4\\2 & 1 & 4 & 3\end{pmatrix} \quad \begin{pmatrix}1 & 2 & 3 & 4\\4 & 3 & 2 & 1\end{pmatrix}$$

18. How many axes of symmetry has an equilateral triangular lamina (i) in its own plane and (ii) otherwise?

    With each of the axes of symmetry can be associated a finite group of rotations of the triangle into itself. What is the order of each group?
    (Oxford and Cambridge G.C.E. A-Level)

19. In the symmetry group of the square, verify that

    (*a*) $(pr)^2 = e$ (*b*) $pr^2pr = r^3$ (*c*) $r^2pr^3 = pr$ (*d*) $r^3pr^2 = pr^3$

20. Write down the product table for the symmetry group of the rhombus and identify the subgroups.

21. Find the symmetry group of the regular pentagon ABCDE denoting a rotation of 72° by r and a half-turn about the axis of symmetry through A by p. Verify that each element is of order 2 and that

    $$r^tp = pr^{5-t} \quad (t = 1, 2, 3, 4)$$

22. Obtain the product table for the symmetry group of the regular hexagon expressing the elements in terms of r, a rotation through 60°, and p, a reflection in an axis of symmetry. Show that

    (*a*) $rpr = p$ (*b*) $(pr^2)^2 = e$ (*c*) $pr^3p = r^3$

    and find the inverses and the orders of all the elements.

8.12 *Subgroups*

Examples of subgroups have already arisen in previous sections. Subgroups will now be considered formally in some detail.

DEFINITION 8.3
A subset H of a group G is a *subgroup* of G provided that H is a group under the binary operation in G.

It should be noted that a subset of a group G is not necessarily a subgroup of G, e.g. the set of integers, positive and negative, together with zero form a group under addition but the subset of positive integers does *not* form a group under addition because it does not contain an identity element.

The above definition requires that, under the binary operation of G,

(i) H is closed so that for any elements $h_1, h_2 \in H$, $h_1h_2 \in H$
(ii) any element $h \in H$ has an inverse $h^{-1} \in H$.

The associative law is automatically satisfied in H since it is satisfied in the group G.

Also, (i) and (ii) imply that the identity element of G is in H since for any $h \in H$, $h^{-1} \in H$ by (ii) and hence $hh^{-1} = e \in H$ by (i).

The following theorem provides a practical criterion for deciding whether a subset of a given group is a subgroup.

THEOREM 8.4
H is a subgroup of G if and only if $h_1h_2^{-1} \in H$ for all elements h_1, $h_2 \in H$.

The condition is necessary since if H is a subgroup, we have

$h_2^{-1} \in H$ for all $h_2 \in H$ by (ii)

Therefore $\quad h_1h_2^{-1} \in H$ by (i)

Conversely, assuming that $h_1h_2^{-1} \in H$ for all h_1, $h_2 \in H$, we have

$h_1h_1^{-1} \in H$ for all $h_1 \in H \qquad$ therefore $e \in H$

Therefore by hypothesis, for all $h_2 \in H$, $eh_2^{-1} \in H$, i.e. $h_2^{-1} \in H$ which verifies (ii)

Hence $h_1(h_2^{-1})^{-1} \in H$, i.e. $h_1h_2 \in H$ which verifies (i)

Therefore H is a subgroup of G.

Any group contains at least two subgroups, the whole group itself and the trivial group consisting of the identity element e only. Any other subgroup is called a *proper* subgroup.

8.13 The Centre of a Group

DEFINITION 8.4
The subset of a group, comprising those elements which commute with all elements of the group, is called the *centre* of that group.

If a group is Abelian, the centre is the whole group. As an example of a non-Abelian group, consider the symmetry group of the equilateral triangle (Section 8.11). The powers of r commute with each other but none commutes with p. Indeed, only one element e commutes with all elements of this group so that its centre is e, the trivial group. In a non-Abelian group, the centre may be a proper subgroup as the following theorem shows.

THEOREM 17.5
The centre of a group is a subgroup.

Let c_1, c_2 be elements in the centre and let g be any element of the group. Therefore

$$c_1g = gc_1 \quad \text{and} \quad c_2g = gc_2$$

$$c_1c_2g = c_1gc_2 = gc_1c_2 \quad \text{thus } c_1c_2 \text{ is an element in the centre}$$

Clearly e is an element in the centre since it commutes with all elements of the group.

Again, for any g, we have

$$c_1g^{-1} = g^{-1}c_1 \qquad (c_1g^{-1})^{-1} = (g^{-1}c_1)^{-1}$$

$$gc_1^{-1} = c_1^{-1}g \quad \text{(see Section 8.5)}$$

Therefore c_1^{-1} is an element in the centre, and thus the centre is a subgroup of G.

8.14 *Cayley's Theorem*

This theorem, due to the British mathematician Cayley (1821–1895), is one of the most fundamental in group theory. It proves that any finite group is isomorphic to a group of permutations.

THEOREM 8.6
Any finite group G, of order n, is isomorphic to a subgroup of the symmetric group S_n of degree n.

Let $g_1, g_2, g_3, \ldots, g_n$ denote the n elements of G. The $n!$ permutations of the numbers $1, 2, 3, \ldots, n$, by which these elements are labelled, comprise the symmetric group S_n.

Let x be a particular element of the group G, then the products xg_1, xg_2, $xg_3, \ldots, xg_n$ will be all different since

$$xg_i = xg_j \Leftrightarrow g_i = g_j$$

and must therefore be a permutation of the elements $g_1, g_2, g_3, \ldots, g_n$.

If we now write

$$xg_1 = g_{i_1}, \qquad xg_2 = g_{i_2}, \qquad xg_3 = g_{i_3}, \ldots, xg_n = g_{i_n}$$

then $i_1, i_2, i_3, \ldots, i_n$ will be a permutation of $1, 2, 3, \ldots, n$.

Now consider the correspondence

$$x \Leftrightarrow X = \begin{pmatrix} 1 & 2 & 3 & \cdots & n \\ i_1 & i_2 & i_3 & \cdots & i_n \end{pmatrix} \tag{8.1}$$

This correspondence associates one permutation X of the group S_n with one particular element x of G. A different element y corresponds to a different permutation since $xg_1 = yg_1 \Leftrightarrow x = y$.

Thus in eqn (8.1), we have a one-to-one correspondence between G and a subset P_n of n permutations belonging to the group S_n.

Let

$$y \Leftrightarrow Y = \begin{pmatrix} 1 & 2 & 3 & \cdots & n \\ j_1 & j_2 & j_3 & \cdots & j_n \end{pmatrix}$$

$$YX = \begin{pmatrix} 1 & 2 & 3 & \cdots & n \\ j_{i_1} & j_{i_2} & j_{i_3} & \cdots & j_{i_n} \end{pmatrix}$$

Now we have

$$xg_1 = g_{i_1} \qquad xg_2 = g_{i_2} \qquad xg_3 = g_{i_3} \qquad \cdots \qquad xg_n = g_{i_n}$$

Therefore

$$yxg_1 = yg_{i_1} = g_{j_{i_1}} \qquad yxg_2 = yg_{i_2} = g_{j_{i_2}} \qquad \cdots \qquad yxg_n = yg_{i_n} = g_{j_{i_n}}$$

$$yx \Leftrightarrow \begin{pmatrix} 1 & 2 & 3 & \cdots & j \\ j_{i_1} & j_{i_2} & j_{i_3} & \cdots & j_{i_n} \end{pmatrix} = YX$$

The one-to-one correspondence (8.1) is thus preserved under the operation of multiplication.

Thus the correspondence is an isomorphism between G and a subset P_n of S_n. Since G is a group, P_n is also a group. G is therefore isomorphic to a subgroup of S_n.

Consider the cyclic group (e, a, b, c) of order 4 whose product table is shown in Table 8.18.

	e	a	b	c
e	e	a	b	c
a	a	b	c	e
b	b	c	e	a
c	c	e	a	b

TABLE 8.18

Let e, a, b, c, be labelled elements 1, 2, 3, 4 respectively. Then

$ae = a$ will be labelled 2, $aa = b$ as 3, $ab = c$ as 4, $ac = e$ as 1

Therefore, in accordance with eqn (8.1),

$$a \Leftrightarrow \begin{pmatrix} 1 & 2 & 3 & 4 \\ 2 & 3 & 4 & 1 \end{pmatrix}$$

Similarly

$$b \Leftrightarrow \begin{pmatrix} 1 & 2 & 3 & 4 \\ 3 & 4 & 1 & 2 \end{pmatrix}$$

$$c \Leftrightarrow \begin{pmatrix} 1 & 2 & 3 & 4 \\ 4 & 1 & 2 & 3 \end{pmatrix}$$

$$e \Leftrightarrow \begin{pmatrix} 1 & 2 & 3 & 4 \\ 1 & 2 & 3 & 4 \end{pmatrix}$$

Thus the cyclic group of order 4 is isomorphic to the subgroup P_4 comprising the above 4 permutations of the symmetric group S_4.

8.15 Lagrange's Theorem

The possible orders of the subgroups of a finite group G are limited by the order of G in accordance with the following theorem.

THEOREM 8.7
If H is a subgroup of order h of a finite group G of order g, then h is a factor of g.

If H is a trivial group, the result is obvious.

Let H be a proper subgroup and let $a_1, a_2, a_3, \ldots, a_h$ be its h ($<g$) distinct elements of which one is e. Let $x \in G$ such that $x \notin H$. Consider the elements

$$xa_1, xa_2, xa_3, \ldots, xa_h \tag{8.2}$$

The elements (8.2) will all be different since, otherwise, if $xa_i = xa_j$ for some i, j, then $a_i = a_j$, which is a contradiction since a_i, a_j are distinct elements of H.

Also no element of the set (8.2) is in H since, otherwise $xa_i = a_j$ for some i, j, i.e. $x = a_j a_i^{-1} \in H$, which is contrary to hypothesis.

The set of elements (8.2) is denoted by xH and is called the *left coset** of G relative to H.

The sets H and xH thus comprise $2h$ elements of G, h of which are in H and h in xH. If G has additional elements, let y be one of these elements such that $y \in G$ but $y \notin H$ and $y \notin xH$.

The left coset yH contains h distinct elements, none of which is in H. Also none is in xH since, otherwise, if $ya_i = xa_j$ for some i, j, then $y = xa_j a_i^{-1}$, i.e. $y \in xH$, which is contrary to hypothesis.

We now have $3h$ distinct elements of G, h of them being in each of H, xH, yH. This process of dividing G into left cosets may be continued indefinitely until the finite number of elements in G is exhausted so that eventually G is partitioned into r sets each of h elements. Therefore $g = rh$ which proves Lagrange's Theorem.

The following theorems may be easily deduced from Lagrange's Theorem.

THEOREM 8.8

If G is a finite group of order p where p is prime, G has no proper subgroups.

p and 1 are the only factors of p. Therefore by Lagrange's Theorem, the only subgroups of G have orders 1 and p. Therefore G has no proper subgroups.

THEOREM 8.9

If x is an element of order r of a finite group G of order n, then r is a factor of n and $x^n = e$.

By Theorem 8.1, $(e, x, x^2, \ldots, x^{r-1})$ form a cyclic group of order r which is a subgroup of G. Therefore r is a factor of n.

Also $x^r = e$ and $n = kr$. Thus

$$x^n = x^{kr} = (x^r)^k = e$$

THEOREM 8.10

The only group of order p, where p is a prime number, is the cyclic group of order p.

By Theorem 8.9 the order of any element $x \neq e$ of the group must be p since p is prime. Thus x generates a cyclic group of order p which must be the whole group. It follows that the only groups of orders 2, 3, 5, 7 are cyclic groups.

* The set $(a_1x, a_2x, a_3x, \ldots, a_hx)$ is called the *right coset* of G relative to H and is denoted by Hx.

THEOREM 8.11
If n is composite, the cyclic group C_n has a unique cyclic group C_r for each factor r of n.

Let x be a generator of C_n so that $x^n = e$. Let $n = rm$. Consider the elements

$$e, x^m, x^{2m}, \ldots, x^{(r-1)m} \quad \text{with } x^{rm} = e$$

These elements are distinct and form a cyclic subgroup of C_n of order r generated by x^m. Thus a cyclic subgroup of order r exists.

It is now necessary to show that this subgroup is unique. Let y be any element of a subgroup of order r.

By Theorem 8.9,

$$y^r = e$$

Let $y = x^k$ where x is a generator of C_n.
Therefore, $y^r = x^{kr} = e$.
Thus kr is a multiple of n. Let $kr = \mu n = \mu rm$.
Therefore, $k = \mu m$. Thus $y = x^k = x^{\mu m}$.

Thus each element of the subgroup of order r is of the form $x^{\mu m}$. Since there are only r distinct elements of this kind, they comprise the whole of the cyclic group of order r generated by x^m.

Thus the cyclic subgroup of order r is unique.

EXAMPLE 8.3
(A) The group of residues (1, 3, 5, 7) (mod 8) under multiplication has a subgroup $H = (1, 5)$. Write down the cosets $H1$, $H3$, $H5$, $H7$.

Solution The group table is given in Table 8.19

	1	3	5	7
1	1	3	5	7
3	3	1	7	5
5	5	7	1	3
7	7	5	3	1

TABLE 8.19

For $H = (1, 5)$,

coset $H1 = (1, 5)$ $(= 1H)$ coset $H3 = (3, 7)$ $(= 3H)$

coset $H5 = (5, 1)$ $(= 5H)$ coset $H7 = (7, 3)$ $(= 7H)$

In accordance with Lagrange's Theorem, the group is partitioned into 4/2, i.e. 2 sets.

(B) H is the subgroup (e, r, r^2, r^3) of the octic group whose table is given in Table 8.17 (p. 246). Show that all the left and right cosets of the octic group relative to H comprise either (e, r, r^2r^3) or (p, pr, pr^2, pr^3). Find the left and right cosets of the group relative to the subgroup $K = (e, p)$.

Solution $H = (e, r, r^2, r^3)$

$$eH = (e, r, r^2, r^3) = He \qquad pH = (p, pr, pr^2, pr^3) = Hp$$

$$eH = He = rH = Hr = r^2H = Hr^2 = r^3H = Hr^3$$

$$pH = Hp = prH = Hpr = pr^2H = Hpr^2 = pr^3H = Hpr^3$$

There are 8/4, i.e. 2 distinct sets.

For the cosets relative to K,

$$\begin{aligned} eK &= (e, p) & pK &= (p, e) \\ rK &= (r, pr^3) & prK &= (pr, r^3) \\ r^2K &= (r^2, pr^2) & pr^2K &= (pr^2, r^2) \\ r^3K &= (r^3, pr) & pr^3K &= (pr^3, r) \end{aligned}$$

$$\begin{aligned} Ke &= (e, p) & Kp &= (p, e) \\ Kr &= (r, pr) & Kpr &= (pr, r) \\ Kr^2 &= (r^2, pr^2) & Kpr^2 &= (pr^2, r^2) \\ Kr^3 &= (r^3, pr^3) & Kpr^3 &= (pr^3, r^3) \end{aligned}$$

In accordance with Lagrange's theorem there are 8/2, i.e. 4 sets in each case.

EXERCISES 8.3

1. Find the centres of the dihedral groups of orders 6 and 8. Show that they contain Abelian subgroups larger than the centre.
[e, (e, r^2), e.g. subgroups in the top left-hand corners of Tables 8.14 and 8.17]
2. Carry out the proof of Cayley's theorem with specific reference to the group for which the multiplication table is

	e	a	b
e	e	a	b
a	a	b	e
b	b	e	a

3. G is the cyclic group $(1, \omega, \omega^2)$ $(\omega = e^{i2\pi/3})$, the operation being ordinary multiplication, and H is the group of residue classes (mod 3) under addition. Show that G and H are isomorphic. Find a permutation group isomorphic to G and H.

$$\left[(1, \omega, \omega^2), (0, 1, 2); \begin{pmatrix}1 & 2 & 3\\1 & 2 & 3\end{pmatrix}, \begin{pmatrix}1 & 2 & 3\\2 & 3 & 1\end{pmatrix}, \begin{pmatrix}1 & 2 & 3\\3 & 1 & 2\end{pmatrix}\right]$$

4. Which of the following subsets are subgroups?
   (*a*) The rationals in the group of real numbers under addition.
   (*b*) The integers, positive and negative (excluding zero) in the group of real numbers under addition.
   (*c*) (i) The rationals and (ii) the integers in the group of real numbers (excluding zero) under multiplication.
   (*d*) The integers, positive, negative and zero, which are multiples of an integer n in the group of real numbers, under addition.
   [(*b*) and (*c* ii) are not subgroups]
5. Show that (1, 5) forms a proper subgroup of the group of residue classes (1, 5, 7, 11) (mod 12) under multiplication. Find another proper subgroup. [(1, 7) and (1, 11)]
6. Show that (1, 4, 13, 16) forms a proper subgroup of the group of the non-zero residue classes (mod 17) under multiplication. Find a subgroup of the subgroup (1, 4, 13, 16). Is there another subgroup of the main group? [(1, 16) is a subgroup of (1, 4, 13, 16) and also of the main group]
7. Identify the subgroups of (*a*) the Vierergruppe (*b*) the cyclic group of order 12. [(*a*) (e, a), (e, b), (e, c); (*b*) subgroups are generated by x^2 (order 6), x^3 (order 4), x^4 (order 3) and x^6 (order 2)]
8. Show that $H = (1, 9, 11)$ is a subgroup of the group of residues (1, 3, 5, 9, 11, 13) (mod 14) under multiplication. Write down the cosets $3H$, $5H$ and $13H$ in each case. [3, 5 and 13]
9. Write down the right and left cosets of
   (i) The Vierergruppe relative to the subgroup (e, a).
   (ii) The dihedral group of order 6 relative to the subgroup (e, r, r^2).
   (iii) The cyclic group $(e, x, x^2, \ldots, x^{11})$ $(x^{12} = e)$ relative to the subgroup (e, x^4, x^8).
   (iv) The symmetric group of degree 4 relative to the subgroup comprising the four permutations given in Exercises 8.2, No. 17.

[(i) (e, a), (b, c), (ii) either (e, r, r^2) or (p, pr, pr^2), (iii) (e, x^4, x^8), (x, x^5, x^9), (x^2, x^6, x^{10}), (x^3, x^7, x^{11}), (iv) As an example, the cosets for $\lambda = \begin{pmatrix}1 & 2 & 3 & 4\\4 & 1 & 2 & 3\end{pmatrix}$ are $\lambda H = \begin{pmatrix}1 & 2 & 3 & 4\\4 & 1 & 2 & 3\end{pmatrix}$ $\begin{pmatrix}1 & 2 & 3 & 4\\2 & 3 & 4 & 1\end{pmatrix}$ $\begin{pmatrix}1 & 2 & 3 & 4\\1 & 4 & 3 & 2\end{pmatrix}$ $\begin{pmatrix}1 & 2 & 3 & 4\\3 & 2 & 1 & 4\end{pmatrix}$]

Index